高等职业教育"十三五"精品规划教材

飞机维修专业英语教程—— 飞机主要结构与部件

主　编　赵迎春　陈凯军
副主编　魏　敏　郭　俊　罗　娜　都昌兵
主　审　杨国余

中国水利水电出版社
www.waterpub.com.cn
·北京·

内 容 提 要

本书由三大模块组成：模块一为飞机维修岗位描述，模块二为飞机维修手册（AMM）介绍，模块三为民用飞机主要结构与部件介绍。全书共分为 12 课。模块一对民航飞机维修岗位具体要求进行详尽的英文阐述。模块二通过对全英文 AMM 的介绍，让学生熟悉飞机维修岗位上高频使用的英文资料。模块三首先介绍了波音 737 飞机五大结构——机身、大翼、起落架、动力装置与尾翼，然后依托这五大结构，分别对尺寸与区域、驾驶舱的主要面板、门的操作、电子舱、起落架、动力装置、辅助动力装置、大翼、飞行操纵面 9 个学习情境进行了专业的英文介绍。这些学习情境结合飞机维修专业材料，通过听、说、读、写、译的语言训练，使学生不仅能够轻易掌握飞机维修的主要英文术语、专业词汇、缩略语，还能熟悉各种英文手册中出现的高频句型与难句，并了解飞机维修岗位的行业规范、要求与 AMM 的编排体系。

本书既可以作为航空类本科与高职院校飞机维修专业及相近专业学生的专业英语教材，也可以作为航空公司的机务工程部及飞机维修公司在职人员的专业英语培训和自学教材。

图书在版编目（CIP）数据

飞机维修专业英语教程. 飞机主要结构与部件 / 赵迎春，陈凯军主编. -- 北京：中国水利水电出版社，2018.8（2023.7 重印）
高等职业教育"十三五"精品规划教材
ISBN 978-7-5170-6724-5

Ⅰ. ①飞… Ⅱ. ①赵… ②陈… Ⅲ. ①飞机－维修－英语－高等职业教育－教材 Ⅳ. ①V267

中国版本图书馆CIP数据核字(2018)第185605号

策划编辑：周益丹　责任编辑：张玉玲　加工编辑：张溯源　封面设计：梁　燕

书　　名	高等职业教育"十三五"精品规划教材 飞机维修专业英语教程——飞机主要结构与部件 FEIJI WEIXIU ZHUANYE YINGYU JIAOCHENG——FEIJI ZHUYAO JIEGOU YU BUJIAN
作　　者	主　编　赵迎春　陈凯军 副主编　魏　敏　郭　俊　罗　娜　都昌兵 主　审　杨国余
出版发行	中国水利水电出版社 （北京市海淀区玉渊潭南路 1 号 D 座 100038） 网址：www.waterpub.com.cn E-mail：mchannel@263.net（答疑） 　　　　sales@mwr.gov.cn 电话：（010）68545888（营销中心）、82562819（组稿）
经　　售	北京科水图书销售有限公司 电话：（010）68545874、63202643 全国各地新华书店和相关出版物销售网点
排　　版	北京万水电子信息有限公司
印　　刷	三河市德贤弘印务有限公司
规　　格	184mm×260mm　16 开本　12.75 印张　307 千字
版　　次	2018 年 8 月第 1 版　2023 年 7 月第 7 次印刷
印　　数	15001—18000 册
定　　价	38.00 元

前　　言

近年来，随着我国民航事业的迅猛发展与民航维修标准的国际化，业内对飞机维修人员的英语阅读水平、翻译水平和交际能力提出了更高的要求。而目前很多航空公司、机场及维修公司等在职机务人员的专业英语水平与综合能力有待提高，部分机务人员虽能看懂维修英文资料，但口头交际和书面沟通能力明显不足。

我国机务人才培养基地主要在各级航空类院校。目前，所有学院都已经意识到飞机维修专业英语的重要性，也认识到了合适的专业英语教材是提升学生英语水平和能力的重要基础。遗憾的是，目前市场上合适的专业英语教材种类并不多。部分教材重视航空方面的阅读文章，虽对提升英语阅读能力有一定的作用，但其实践性与岗位针对性不强，且缺乏听、说、读、写、译能力的系统训练，不能满足岗位对英语综合能力的要求。为此，有些院校节选飞机维修手册部分章节的复印资料作为教材，尽管实用性强，却缺乏理论基础，更没有配套练习。同时，市场上大部分教材着重文字表述，缺少实物插图，再加上航空类专业英语词汇与术语非常抽象，涉及机械、通信、电子、电源、液压等诸多专业领域，这使得专业英语教学达不到理想的效果。

根据机务工作的实际需求和情境，针对飞机维修及相关专业对英语的个性化需要，参照中国民航局颁发的《CCAR-147-02 部民用航空器维修基础培训大纲》、CCAR-66 部《维修人员培训大纲》及各航空公司相关岗位的实际需求与维修工作中的工作程序，借鉴飞机维修手册的部分内容，再结合飞机维修的其他英文资料，配以实物插图，编写了这本集听、说、读、写、译能力为一体的《飞机维修专业英语教程——飞机主要结构与部件》。本书继承了传统教材注重英文阅读与翻译能力培养的特点，同时加大了对航空专业领域内英语听力、口语表达及英文写作的培养力度，具有如下鲜明的特色。

1. 基于飞机维修职业能力需求开发学习模块

本书从深入了解岗位能力需求，到熟悉飞机维修工作中常用的英文资料，再到学习飞行器的主要部件，全书自始至终围绕"职业能力培养"这个重心，针对飞机维修岗位对英语的特殊要求开发专业英语的学习模块。

2. 依托飞机五大结构编排学习情境

本书基于阶梯学习模式，在熟悉岗位需求与全英文维修资料之后，再依托飞机五大结构上的主要部件来编排学习情境，条理清晰，具有内在逻辑性。通过学习这些情境的英文术语和高频句型，循序渐进地使学生熟练两方面的内容：①掌握飞机机内、机外主要部件上出现的英文标识，包括英文专业术语、大写缩略语和英文句型；②熟悉全英文维修资料中出现的英文专业词汇、术语、缩略语、高频句型及难句。

3. 按照英语听、说、读、写、译五大能力细化专业学习内容

每一个学习情境的内容由与飞机维修相关的专业听说、专业阅读、专业翻译、专业写作四大部分组成。专业听说帮助即将进入飞机维修领域的学生进一步自我提升，为更好地与国

外技术人员交流而设计；专业阅读与专业翻译使学生熟悉飞机主要的结构及零部件中出现的各种英文术语、高频词汇、缩略语、高频句型和难句，以及工作中出现的各种英文文献，使学生具备读懂全英文手册的能力；专业写作是为了让学生在维修工作中能够用英文做出维修故障描述、维修故障诊断并提出维修故障建议而设计。

4. 糅合英语学习与岗位需求，体现无缝对接

本书系统介绍了民航维修的具体需求，详细介绍了飞行器五大结构的主要部件，结合AMM手册，对飞行器及飞机维修英文资料中出现的主要英文专业词汇、术语、各种缩略语和高频句型进行介绍，使学生能以英语为媒介进行专业学习，以达到获得专业知识的目的。本书中的所有材料皆来自于实际工作中所用到的维修资料，所有练习都是为了让学生能够在实际工作中对所遇到的英语专业资料轻松驾驭，相关视频资料也是来自国际和国内一线工作的全英文视频。

5. 专业阅读配以实物图解，体现直观生动

本书的阅读文章通过专业讲解与实物插图结合，图文并茂地将飞行器主要部件展现出来，生动形象，使学生能轻易把抽象的英文术语与实物相联系，从而使专业术语的学习更加直观和高效。

6. 配套立体化资源，拓展全方位学习渠道

本书为学生提供了多渠道、多元化的学习资料，涵盖了电子教案、PPT、飞行器部件图片库、闯关卡、微课等各种资料，结合当前技术水平和现实条件基础，教材的内容、形式、服务都可通过扫码获取，最大限度地满足学生学习飞机维修专业英语的个性化需求，打破了以往传统英语教材形式的单一性，提高了其实用性，满足现代学生个性化、自主学习的要求。

本书可作为本科类和高职类航空院校的飞机维修、发动机维修专业的专业英语教材，也可作为飞机电子设备维修、通用航空、飞机维修（士官方向）、无人机飞行器、理化测试与质检等专业向民航拓展的专业英语资料，亦可作为所有民航在职机务人员、致力于向民航拓展的航修人员的专业英语自学材料。本书可与《飞机维修专业英语教程——飞机系统》共同作为系列教材使用，亦可单独使用。

本书由长沙航空职业技术学院赵迎春副教授、陈凯军副教授任主编，中国南方航空公司机务培训中心质量经理、全国发动机维修技能大赛裁判长杨国余先生担任主审，由魏敏、郭俊、罗娜、都昌兵任副主编，李娜、王志敏、王江、唐启东、徐乐及闵雅婷参与了教材编写。学术委员会主任 & 教材委员会主任关云飞、主管教学副院长朱国军、基础教育学院院长唐倩、机电学院院长熊纯、CCAR147机务培训中心质量经理江游、教务处处长彭圣文对本教材给予了大量技术性支持及帮助。在此，对本教材的编写给予指导、支持、帮助的领导、专业英语教研室同事、飞机教研室同事，以及所有提供过帮助的学生表示真诚的感谢。在编写过程中，本书参考了大量的书籍、资料和各种型号的飞机维修手册，在这里谨向本书提供帮助的机务维修一线的领导和技术人员表示诚挚的感谢。

本书配有听力录音、课文朗读录音、课文译文、课文配套微课、习题答案等资源，学习者可以扫描书中二维码查看更多内容。中国慕课平台上开设了在线课程，有大量配套教学资源及补充资源，链接地址为 https://www.icourse163.org/course/CAVTC--1461652166，课程名称

是"航空维修专业英语"，主持人为赵迎春，团队成员有魏敏、郭俊、王江、王志敏、李娜。

由于编者实践经验有限，加之时间仓促，书中难免存在疏漏甚至错误之处，恳请广大读者和专家批评指正，编者会在本书的使用过程中仔细检查，并一一校正。

扫码看慕课

编　者

2018 年 6 月

Contents

Module One Post Description and Requirements for Civil Aircraft Maintenance Personnel

Introduction

Many college students have no much idea about what an aircraft mechanic is and what qualifications are required on the post, not knowing what to do and how to do it.

This module gives a detailed description of the post, and students will know what an aircraft mechanic job is like and what an aircraft mechanic needs to prepare for FAA(Federal Aviation Administration) Certificate. And they will understand the importance of professional English for an aviation mechanic.

The pictures, as shown from Fig.M1-1 to Fig.M1-6, will help you to know how to become an aircraft mechanic.

Fig. M1-1　Aircraft Maintenance Training Shop in CAVTC
(Changsha Aeronautical Vocational & Technical College)

Fig.M1-2　CCAR Part 147 Skill Training

Fig. M1-3　Drills with Sichuan Airline

Fig. M1-4　Aircraft Maintenance Majors in CAVTC

Fig. M1-5　Fastener & Safety Training

Fig. M1-6　Sheet Metal Training

Lesson 1 Post Description of a Civil Aircraft Mechanic

Learning Objectives:

1. Knowledge objectives:

　　A. To master the major words, related terms and abbreviations about aircraft maintenance.

　　B. To master the key sentences.

　　C. To know the duties and responsibilities of civil aircraft mechanics.

2. Competence objectives:

　　A. To be able to understand frequently-used & complex sentence patterns, acronyms and obtain key information on aviation maintenance quickly.

　　B. To be able to talk about aviation or aircraft with English speakers.

C. To be able to fill in job cards in English.

3. Quality objectives:

To be able to self-study with the help of aviation dictionaries, Internet or other resources.

Section 1 Aviation Listening and Speaking

听力录音

1.1 Aviation Listening: listen to the audio and fill in the blanks with the missing words.

I'll explain how you become an 1. _____ captain! Now let's imagine you've been 2. _____ for an airline for quite some time, you've gained 4000-5000 flying 3._____ over the past 7-8 years. You've not only profited from your flight experience, you've also gained seniority over the years which is a crucial part in becoming a 4._____.

Now your airline makes an announcement for a so called "Captains Upgrading". There can be many reasons why the airline is in need of new captains. Older captains might have retired or left the company for other reasons, or the company is an overall growth, more planes need more 5._____.

So you apply for the captains upgrade if you're one of the 6._____ with a higher seniority. Sooner or later the department of flight training will set you up for the PICUS (Pilot In Command Under Supervision) program, meaning you as a co-pilot act as captain from the right hand seat. You'll be the pilot in command, fulfilling captain duties, like passengers announcements, dealing with situations with ground handling, 7._____ to the 8._____ and much more.

The amount of PICUS hours vary for many different reasons, for example in which country you're flying, flight 9._____ 10._____ , aircraft type, previous gained hours, etc.

1.2 Aviation Speaking: look at the pictures below and describe them in details.

Clues: aircraft mechanic; aircraft parking; civil; post.

Section 2 Aviation Reading

课文朗读录音　　译文

Pre-reading questions:

1. Do you have the dream of becoming an aircraft mechanic?

2. Do you know what they have to do on the post?

Post Description of a Civil Aircraft Mechanic

An aircraft mechanic is responsible for the maintenance, repair and the overhaul(MRO) of an aircraft . Graduates who have got the necessary work experience and obtained their license will be qualified to sign off on the airworthiness of an aircraft. Employment opportunities will be provided by airlines and MRO centres. Graduates with the license can earn up to RMB 10,000 a month or even more while those without can only earn about RMB 6,000.

Fig.1-1　Work of Aircraft Mechanics

Civil aircraft mechanics is a particular job with many strict rules and requirements. According to the requirements of Civil Aviation Administration of China (CAAC), only those personnel who have got practical skill training and passed the examination could be eligible for independent maintenance and apply for the basic part of the maintenance personnel licencing. The purpose is to improve basic competence of maintenance personnel so as to reduce maintenance error and keep flight safety. One of the most important skills to become a mechanic is to read and comprehend different kinds of English manuals. Technical manuals will teach you what you need to know.

What is an Aircraft Mechanic Job Like?

Fig.1-2　Maintenance Work in Apron & Hangar

Aircraft mechanics work in hangars, on the field, on the "flight lines" where aircraft park, or in repair stations. They use hand and power tools as well as sophisticated test equipment. Maintenance is performed around the clock, seven days a week. New mechanics and technicians have to work at night and on weekends. The noise level both indoors and on the flight line could be very high. Sometimes the work requires physical activity from climbing ladders to crawling. They work under pressure to make sure an airplane fly on time.

What does an Aircraft Mechanic Need for a FAA Certificate?

微课

Fig.1-3　Textbooks of FAA Certificate I

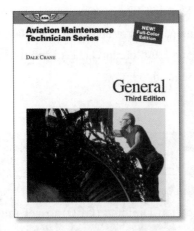 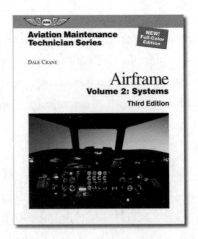

Fig.1-4　Textbooks of FAA Certificate II

> **Does an Aircraft Mechanic Need a License?**

Not necessarily.

However, if you don't have a mechanic's certificate from the FAA, you can work only supervised by someone who does have a certificate. You cannot approve equipment for return to service. Without a certificate, you are less likely to advance to the top of the career field. The FAA issues mechanics and repairman certificates. Mechanics can get either an airframe rating or a powerplant rating–most mechanics get both. Repairmen get certificates with ratings to perform only specific tasks, and they must be associated with FAA-approved Repair Stations, commercial operators, or air carriers holding authority to perform these tasks.

> **Does an Aircraft Mechanic Need Any Experience to Become a Certified Mechanic?**

Yes. An aircraft mechanic can get the required experience through civilian or military on-the-job training, or by attending a special school for aircraft mechanics.

> **Does an Aircraft Mechanic Have to Take Any Tests?**

Yes. An aircraft mechanic has to take both oral and practical tests. Fee needs to be paid for each test. The oral and practical tests cover 43 technical subjects. Typically, tests for one certificate–airframe or powerplant–last about 8 hours.

Requirements to Get a FAA Certificate

There are basic requirements; experience requirements; oral, practical and written test requirements to become an aircraft mechanic.

Basic Requirements

To get a repairman's certificate, you must be recommended by a repair station, commercial operator, or air carrier. You must

> be at least 18 years old.

> be able to read, write, speak, and understand English.

> be qualified to perform maintenance on aircraft or components.

> be employed or do a specific job requiring special qualifications by an FAA-certified

Repair Station, commercial operator, or air carrier.

> ➢ be recommended for the repairman certificate by your employer.
> ➢ have either 18 months practical experience in the specific job or complete a formal training course acceptable to FAA.

Experience Requirements to Become an Aircraft Mechanic

> ➢ An aircraft mechanic can attend one of the 170 FAR Part147 Aviation Maintenance Technician Schools nationwide. These schools offer training for one mechanic's certificate or both. You must present an official letter from your military employer certifying your length of service, the amount of time you worked in each specialties.

You need a high school diploma or a General Education Diploma (GED) to get into most schools. The schooling lasts from 12 months to 24 months, generally less than required by FAA for on-the-job training. When you graduate, you are qualified to take FAA`s exams. Graduates often get higher starting salaries than individuals who got their required experience in one of the other two ways.

> ➢ You can work at FAA Repair Station or FBO under the supervision of a certified mechanic for 18 months for each certificate, or 30 months for both. You must document your experience with pay receipts, a log book signed by your supervising mechanic, a notarized statement from your employer, or other proof to show you have worked the required time.
> ➢ You can join one of the armed services and get training and experience in aircraft maintenance. Make sure you are in a military occupational specialty for which FAA gives credit. You can get a current list of acceptable specialties from the local FAA Flight Standards District Office (FSDO).

You must present an official letter from your military employer certifying your length of service, the amount of time you worked in each specialties, the make and model of the aircraft or engine on which you got practical experience, and where you got the experience. You cannot count time you spent training for the specialty, only the time you spent working in the specialty.

With both types of on-the-job training you should set aside additional study time to prepare for the written and oral/practical tests. The FAA will give you credit for your practical experience only after they review your paperwork and you have a satisfactory interview with an FAA Airworthiness inspector.

Oral, Practical & Written test

To become an aircraft mechanic, you must take oral and practical tests as well as written tests. There is fee for each test. A Designated Mechanic Examiner gives you the oral and practical tests. You can get a list of these examiners at the local FAA office. The oral and practical tests cover 43 technical subjects. Typical tests for one certificate-airframe or powerplant takes about 8 hours.

To take the written test, you must present your proof of experience to an FAA inspector at the local FAA office. There are separate tests for airframe and powerplant mechanic certificates, as well as a general test covering both. If the inspector decides you meet the requirements to take one of the tests, you may make an appointment for testing at one of the many computer testing facilities (PDF) world wide. You can get a list of sample questions general, airframe, and powerplant test.

If you fail part of a test, you have to wait for 30 days before you can take it again, unless you give a letter to the Examiner showing you've gotten additional training in the areas you failed. You must pass all the tests within a 24-month period. The FAA will then issue you a certificate.

Abbreviations and Acronyms

1. Civil Aviation Administration of China (CAAC)　中国民航局
2. Federal Aviation Administration （FAA）　　美国联邦航空管理局

Words and Phrases

1.	aircraft mechanic	飞机机械工
2.	maintenance *n.*	维护，维修
3.	overhaul *n.*	大修
4.	licence *n.*	执照
5.	airworthiness *n.*	适航
6.	aircraft maintenance personnel license	维修人员执照
7.	hangar *n.*	机库
8.	rating *n.*	执照，等级
9.	eligible *adj.*	合格的
10.	mechanic *n.*	机械工
11.	sophisticated *adj.*	复杂的
12.	supervise *v.*	监督
13.	certificate *n.*	证书
14.	airframe *n.*	机身
15.	powerplant	动力装置
16.	repair station	维修站
17.	commercial operator	商业运营商
18.	air carrier	航空承运人
19.	authority *n.*	权威
20.	civilian *adj.*	民用的
21.	military *adj.*	军事的
22.	nationwide *adj.*	全国的
23.	diploma *n.*	证书
24.	document *n.*	文件
25.	receipt *n.*	收据
26.	certify *vt.*	证明
27.	specialty *n.*	专业
28.	airworthiness inspector	适航检查员

29. log book   日志
30. on-the-job   在职的

Exercises

2.1 Answer the following questions according to the passage.

1. Who can maintain the aircraft on the independent basis ?

2. How many licenses do you need to be a mechanic?

3. How are these licenses available?

4. Does an aircraft mechanic need any experiences to become a certified mechanic?

5. Does an aircraft mechanic have to take any tests?

2.2 Translate the following terms or abbreviations into Chinese.

1. airworthiness   2. mechanic

3. powerplant   4. repair station

5. air carriers   6. hangar

7. airframe rating   8. powerplant rating

9. airworthiness inspector   10. flight standard district

2.3 Give the equivalent English terms and corresponding Chinese translations according to the pictures.

1.＿＿＿＿＿

2.＿＿＿＿＿

3.＿＿＿＿＿

4.＿＿＿＿＿

5._____

6._____

Section 3 Aviation Translation

Translate the following English sentences into Chinese.

1. An aircraft mechanic is responsible for the maintenance, repair and the overhaul(MRO) of an aircraft.

2. Repairmen get certificates with ratings to perform only specific tasks, and they must be associated with FAA-approved Repair Stations, commercial operators, or air carriers holding authority to perform these tasks.

3. One of the most important skills to become a mechanic is the ability to read and comprehend different kinds of English manuals.

4. To become an aircraft mechanic, you must take oral and practical tests as well as written tests.

5. You must pass all the tests within a 24-month period. The FAA will then issue you a certificate.

Section 4 Aviation Writing

Situations: Yafeng Guo is an aircraft mechanic. He is inspecting the plane and finding some faults. Please help him write down fault description. Some hints of the description words, phrases & terms and the key sentences about fault location, fault description, fault solutions are offered as follows.

Key words, phrases & terms:

1. 左 L / LH
2. 右 R / RH
3. 前部 forward/fwd
4. 后部 afterward /aft
5. 上面 upper
6. 下面 lower
7. 左上 upper left
8. 右下 lower right
9. 左前 left forward

10. 右后 right afterward

11. 内侧 inboard / I/B

12. 外侧 outboard / O/B

13. 左内侧 left inboard

Key sentences of direction description:

1. 检查发现左后航行灯不亮。L AFT position light is not on after checking.

2. 检查发现右前轮磨损超标。R tire of the nose landing gear is worn out of limits.

3. 向左 / 右偏。Pulled to the left/right.

4. 检查发现主起落架左外刹车组件磨损超标。

   L outboard wear indicator pin of MLG brake assy is out of limits.

5. 更换左外起落架轮子。Replace L/H outboard main wheel.

Task: Yafeng Guo finds out three problems in the checking. First, the right position light can not work, Second, the right tire of the nose landing gear is worn out of limits, and the last, replace right outboard main wheel needs to be replaced. Please finish the writing task.

Module Two　Aircraft Maintenance Manual

Introduction

There are tremendous English documents on aircraft repair and maintenance. Different maintenance documents provide help for all maintenance activities. They work together to help an aircraft maintenance mechanic do the scheduled maintenance and unscheduled work, including Aircraft Maintenance Manual (AMM), Maintenance Planning Documents (MPD), Fault Isolation Manual (FIM),Built-In Test Equipment (BITE) Manual, Structural Repair Manual (SRM), Dispatch Deviations Guide (DDG), etc. Aircraft Maintenance Manual(AMM) is developed to have the latest technical data which contains the information necessary to service, troubleshoot, functionally check and repair. It includes all installed items and equipment that normally requires service on the flight line or maintenance hangar. All chapters of AMM are arranged according to Air Transport Association Specification (ATA-100).

In this module AMM of Boeing is taken to illustrate the details.

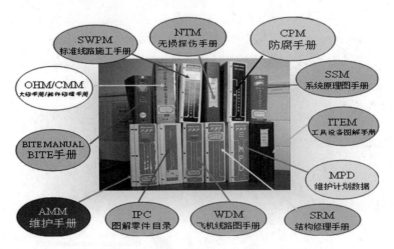

Fig. M2-1 Different Manuals and Documents

Lesson 2　Introduction of Aircraft Maintenance Manual

Learning Objectives:

1. Knowledge objectives:

A. To master the major words, related terms and abbreviations about manuals.

B. To master the key sentences.

C. To master the structures of AMM.

2. Competence objectives:

A. To be able to understand frequently-used & complex sentence patterns, acronyms and obtain key information on aviation maintenance in AMM.

B. To be able to talk about different manuals in English.

C. To be able to search information in AMM and fill in job cards in English.

3. Quality objectives:

To be able to self-study with the help of aviation dictionaries, Internet or other resources.

Section 1 Aviation Listening and Speaking

听力录音

1.1 Aviation Listening: listen to the audio and fill in the blanks with the missing words.

In this audio I'll 1._____ how to fill out the pilot's 2._____. The pilot's logbook is a 3._____ element in your 4._____ on becoming an airline pilot. If you want to 5._____ any flight job, the 6._____ will ask you how many hours you have 7._____, to 8._____ how much flying experience you've had in the past. Nowadays, there are of course all sorts of Apps to record your flight hours, but I highly 9._____ that you stick to old fashion way and write down your hours in a 10._____ pilot's logbook. All the best and have fun collecting your precious first flight hours!

1.2 Aviation Speaking: look at the pictures below and describe them in details.

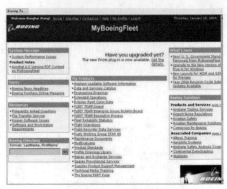

Clues: an aircraft mechanic; help; trouble; solutions; manuals; job card.

Section 2 Aviation Reading

课文朗读录音　　　译文

Pre-reading questions:

1. Where can the aircraft mechanics get help if problems emerge?

2. What do you expect a manual to be like?

Aircraft Maintenance Manual for Boeing 737

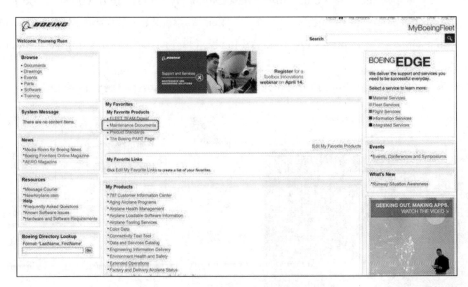

Fig. 2-1　Manuals Online

Different maintenance documents for the Boeing supply help for all maintenance activities. They work together to help do the maintaining work. The maintenance documents such as AMM and MPD help do scheduled maintenance work with SSM,WDM, SRM, IPC,etc. giving supporting data, while FRM, FIM, BITE, SRM, DDG, AMM do unscheduled maintenance work.

The scheduled maintenance work includes:

微课

- ➢ Stop checks
- ➢ Airplane turn around
- ➢ Daily checks
- ➢ Planned checks

Use these documents to do scheduled maintenance:

- ➢ Maintenance Planning Documents (MPD)
- ➢ Airplane Maintenance Manual (AMM)

The following documents supply supporting data to do scheduled maintenance:

- ➢ System Schematics Manual (SSM)
- ➢ Wiring Diagram Manual (WDM)
- ➢ Structural Repair Manual (SRM)
- ➢ Illustrated Parts Catalog (IPC)

The unscheduled maintenance work includes:

- ➢ Flight faults
- ➢ Ground faults
- ➢ Service problems
- ➢ Structural damage

Use these documents to do unscheduled maintenance:

- ➢ Fault Reporting Manual (FRM)
- ➢ Fault Isolation Manual (FIM)
- ➢ Built-In Test Equipment (BITE) Manual
- ➢ Structural Repair Manual (SRM)
- ➢ Dispatch Deviations Guide (DDG)
- ➢ Airplane Maintenance Manual (AMM)

The Relationship between AMM and ATA-100

All chapters of AMM are arranged according to Air Transport Association Specification (ATA-100). In accordance with ATA-100, AMM provides procedures for both scheduled and un-scheduled maintenance. For easy reference, introductory material on the manual is located in the front of the text. This material includes the title page, transmittal letter and temporary revision list, introductory section, chapter summary and service bulletin list.

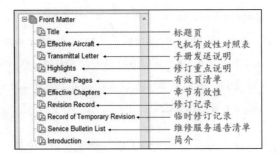

Fig. 2-2　Front Matter

Division of AMM

Aircraft Maintenance Manual (AMM) is divided into two parts:

Part I —Systems Description Section (SDS), SDS replaces the description and operation (D & O) section of 737-300/400/500.

Part II—Practices and Procedures (P&P). P&P contains the maintenance practices and procedures to do maintenance on the airplane.

Both Parts show the configuration of the airplane in an operator's fleet. They have frequent revisions for improvements and for configuration changes.

SDS introduction

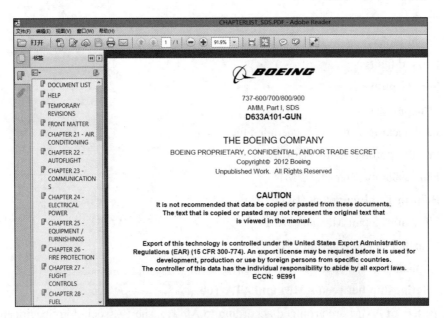

Fig. 2-3　AMM Part I-SDS

SDS gives description of the interfaces, function, configuration and control of the airplane system and subsystems, explains how systems and components operate, shows how systems are constructed, and describes how interfaces relates to other systems. With description you will become familiar with the airplane systems so that you can do fault isolation and system maintenance. The SDS content can be used as a training manual.

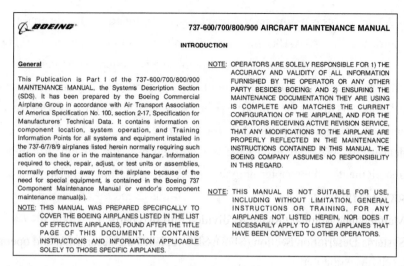

Fig. 2-4　SDS Introduction

SDS organization is by ATA chapter (system) or chapter/section (subsystem). Each subject of SDS has the following information:

➢　Purpose/introduction

➢　General description

➢　Component location

➢　Interface

➢　Operation

➢　Functional description

➢　Training Information Points (TIPS)

P&P introduction

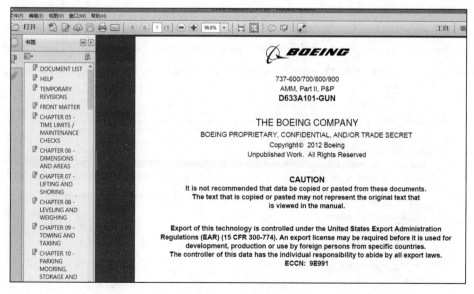

Fig. 2-5　AMM Part II-PP

Part II of AMM has the maintenance practices and procedures to do maintenance on the airplane. In accordance with ATA-100, the manual is divided into chapters and groups of chapters. These represent a functional break-down of the airplane and its systems. Chapters 1 through 4 are for information that customer airline wants to accumulate on a particular airplane, Chapters 5-12 are for Aircraft General, Chapters 20-49 for Aircraft System, Chapters 51-57 for Aircraft Structure, Chapters 60-67 for Propeller/Rotor , and Chapters 70-80 for Powerplant.

In accordance with ATA-100, each page is identified by a three element number and a page number. The first element is the chapter number, representing the functional system of the airplane. The second element identifies a section or sub-system of this chapter. And the third element identifies a subject or the component within the same system. Within each tab section, the manual has more divisions that use the assigned subject number (ASN) list number XX-YY-ZZ, where:

- ➤ XX is the ATA chapter.
- ➤ YY is the subsystem or sub-subsystem.
- ➤ ZZ is the unit (component).

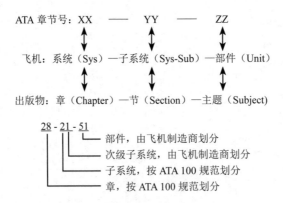

Fig. 2-6　The Meaning of ASN

When the text material is about the chapter in general, the section and the subject number are zeros. When the text material in the section is of general nature, as in this case, the subject numbers are zeros.

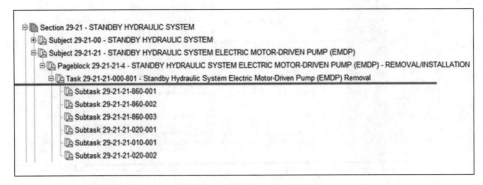

Fig. 2-7　ASN in Contents of AMM

737-600/700/800/900
AIRCRAFT MAINTENANCE MANUAL
FUEL VENT FLOAT VALVE - REMOVAL/INSTALLATION

1. **General**
    A. This procedure has two tasks:
       (1) A task to remove the float valve for the fuel vent system
       (2) A task to install the float valve for the fuel vent system.
    TASK 28-13-11-000-801
2. **Fuel Vent Float Valve Removal**
    (Figure 401)
    A. References

Reference	Title
28-11-00-910-802	Purging and Fuel Tank Entry (P/B 201)
28-11-11-000-801	Main Tank Access Door Removal (P/B 401)

    B. Location Zones

Zone	Area
532	Left Wing - Main Tank, Rib 5 to Rib 22, Wing Station 204.25 to Wing BL 643.50
632	Right Wing - Main Tank, Rib 5 to Rib 22, Wing Station 204.25 to Wing Station 643.50

    C. Access Panels

Number	Name/Location
532RB	Main Tank Access Door - Wing Station 629
632RB	Main Tank Access Door - Wing Station 629

    D. Procedure
    SUBTASK 28-13-11-650-001

    WARNING: DO ALL THE SAFETY PROCEDURES TO DEFUEL THE TANK AND TO GO INTO IT. INJURY TO PERSONS AND DAMAGE TO EQUIPMENT CAN OCCUR IF YOU DO NOT FOLLOW THE SAFETY PROCEDURES.

    (1) Prepare the applicable fuel tank for entry (TASK 28-11-00-910-802).
    SUBTASK 28-13-11-010-001
    (2) Open these access panels:

Number	Name/Location
532RB	Main Tank Access Door - Wing Station 629
632RB	Main Tank Access Door - Wing Station 629

    (TASK 28-11-11-000-801).
    SUBTASK 28-13-11-020-001
    (3) Remove the bolts [2] that attach the float valve [1] to the structure.
    SUBTASK 28-13-11-020-002
    (4) Remove the float valve [1].

    ——————— END OF TASK ———————

    EFFECTIVITY
    GUN ALL

    28-13-11
    Page X
    month/date/year

Fig. 2-8   ASN Demonstration

Boeing publications designates a page code for each page. Each page has two numbers in the lower right corner: the ASN and a topic page number. Each topic is made up of a page block. Each page block is for a special type of information.

The description and operation of the system, components and information is found in the page block from 1 to 100. The topic of troubleshooting is presented in pages 101 to 200. The topic of maintenance practices are assigned to 201 to 300 page block. If they are detailed, they will be located in the sub-topics, pages 301 to 900.

Table 2-1   Page Block Arrangement

Page Type	Page Block
DESCRIPTION AND OPERATION (D&O)	1-99
TROUBLESHOOTING (TS)	101-199
MAINTENANCE PRACTICES (MP)	201-299
SERVICE (SRV)	301-399
REMOVAL/INSTALLMENT (R/I)	401-499
ADJUSTMENT/TEST (A/T)	501-599
INSPECTION/CHECK (I/C)	601-699
CLEANING/PAINTING (C/P)	701-799
APPROVED REPAIRS (AR)	801-899
DISPATCH DEVIATION GUIDE (DDG)	901-999

Effectivity is the method of defining an airplane configuration and showing configuration differences among airplanes. The list of effective pages is a numerical list of all pages effective for the chapter. Pages that have been revised, added or deleted in the latest periodic revision are indicated by an asterisk. At the bottom of each page, there is a space to show airplane effectivity. There are several notations used to indicate effectivity. For example, This page is effective for all airplanes covered by this manual.

②飞机有效性对照表(Effective Aircraft)：

BOEING   737-600/700/800/900 AIRCRAFT MAINTENANCE MANUAL

This manual is applicable to the aircraft in this list:

| Model-Series | Operator | | Manufacturer | | | Registration Number |
	Identification Code	Effectivity Code	Block Number	Serial Number	Line Number	
737-7BK	TCI	001	YB276	30617	812	D-AHIC
737-7BK	TCI	050	YM101	33015	1384	VH-VBV
737-7BK	TCI	051	YM102	33025	1707	VT-SIZ
737-7BK	TCI	052	YM103	33026	1715	VT-SJA
737-81Q	TCI	104	YC904	30618	830	G-OXLD
737-81Q	TCI	105	YC905	30619	856	N733MA
737-8BK	TCI	201	YD251	30620	991	VH-VOA
737-8BK	TCI	202	YD252	30622	1108	VH-VOB

显示适用于该手册的所有飞机的各种编号

Fig. 2-9   Effectivity of AMM

Subject/Page	Date	COC	Subject/Page	Date	COC
HIGHLIGHTS (cont.)			RECORD OF TEMPORARY REVISIONS		
D 38	Oct 15/2008		1	Oct 10/2002	
D 39	Oct 15/2008		2	Oct 10/2002	
D 40	Oct 15/2008		SERVICE BULLETIN LIST		
D 41	Oct 15/2008		1	Oct 10/2003	
D 42	Oct 15/2008		2	BLANK	
D 43	Oct 15/2008		INTRODUCTION		
D 44	Oct 15/2008		1	Feb 10/2005	
D 45	Oct 15/2008		2	Feb 10/2005	
D 46	Oct 15/2008		3	Oct 10/2006	
D 47	Oct 15/2008		4	Oct 10/2002	
D 48	Oct 15/2008		5	Feb 10/2003	
D 49	Oct 15/2008		6	Oct 10/2002	
D 50	BLANK		7	Oct 10/2002	
EFFECTIVE PAGES			8	Jun 10/2003	
1 thru 2	Oct 15/2008		9	Oct 10/2002	
			10	BLANK	
EFFECTIVE CHAPTERS					
O 1	Oct 15/2008				
O 2	Oct 15/2008				
REVISION RECORD					
1	Oct 10/2002				
2	Oct 10/2002				

A = Added, R = Revised, D = Deleted, O = Overflow, C = Customer Originated Change

EFFECTIVE PAGES

A：表示增加页；
R：表示改版页；
D：表示删除页；
O：表示覆盖页；
C：表示应客户要求发起的更改。

Fig. 2-10   Different Revisions for AMM

Servicing is another major subject in AMM. There is an entire chapter devoted to scheduled and unscheduled servicing applicable to the whole airplane. Chapter 12 has the title SERVICING. This chapter has procedures to fill and drain items such as fuel, oil, hydraulic, fluid, water and tire pressure. Servicing information is also located within the other chapters in the manual. The procedures in the 301 to 399 page block relate to the maintenance practices covered in those chapters includes items such as servicing shock struts, lubricating control cables, and sterilizing potable water system.

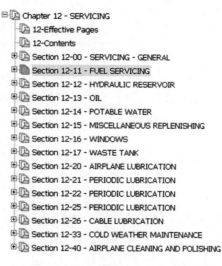

Fig. 2-11　Chapter 12-Servicing in AMM

Abbreviations and Acronyms

1.	Airplane Maintenance Manual (AMM)	飞机维修手册
2.	Assigned Subject Number (ASN)	指定功能号
3.	Air Transport Association (ATA)	航空运输协会
4.	Dispatch Deviation Guide (DDG)	放行偏差指南
5.	Fault Isolation Manual (FIM)	故障隔离手册
6.	Fault Reporting Manual (FRM)	故障报告手册
7.	Illustrated Parts Catalog (IPC)	零件图解目录
8.	Minimum Equipment List (MEL)	最低设备清单
9.	Master Minimum Equipment List (MMEL)	主最低设备清单
10.	Maintenance Planning Document (MPD)	维修计划文献
11.	Practice and Procedure (P&P)	实操与程序
12.	Systems Description Section (SDS)	系统描述部分
13.	Structural Repair Manual (SRM)	结构修理手册
14.	Service (SRV)	勤务
15.	System Schematics Manual (SSM)	系统原理图手册
16.	Wiring Diagram Manual (WDM)	线路图手册

Words and Phrases

1.	troubleshoot *v.*	排除故障（排故）
2.	hangar *n.*	飞机库
3.	break-down *n.*	分类
4.	scheduled maintenance	预定性维修
5.	customize *v.*	定做，定制
6.	configuration *n.*	构型；结构；形态
7.	reference *n.*	参考
8.	title *n.*	标题
9.	temporary revision list	临时修改清单
10.	transmittal letter	手册发送说明
11.	service bulletin	服务通告
12.	asterisk *n.*	星号
13.	notation *n.*	注释，说明
14.	effectivity *n.*	有效性
15.	page code	页码
16.	page block	页块
17.	fault isolation	故障隔离
18.	circuit diagram	电路图
19.	fuel *n.*	燃油
20.	oil *n.*	滑油
21.	hydraulic fluid	液压油
22.	tire pressure	轮胎压力
23.	tank *n.*	油箱
24.	reservoir *n.*	水箱
25.	standard specification	标准规范；标准规格
26.	sterilize *n.*	消毒；杀菌
27.	potable water	饮用水

Exercises

答案

2.1 Answer the following questions according to the passage.

1. What does the aircraft maintenance manual include and what does it do?

2. What does the scheduled maintenance work include?

3. What is AMM based on?

4. What does effectivity refer to?

5. Where can you find the descriptions and operations of the system, components, and information?

2.2 Translate the following terms or abbreviations into Chinese.

1. AMM   2. ATA

3. CL   4. LRU

5. maneuvering control   6. scheduled maintenance

7. effectivity   8. air-cycle pack

9. fault isolation   10. hydraulic fluid

2.3 Give the equivalent English terms and corresponding Chinese translations according to the pictures.

1._____

2._____

3._____

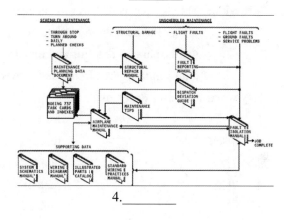

4._____

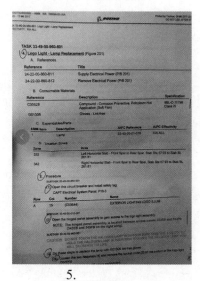

5._____

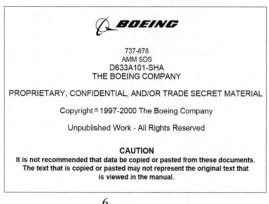

6._____

Section 3 Aviation Translation

Translate the following English sentences into Chinese.

1. Different maintenance documents for the Boeing provide help for all maintenance activities.

2. The maintenance documents such as AMM and MPD help you do scheduled maintenance work with SSM,WDM, SRM, IPC ,etc. giving supporting data, while FRM, FIM, BITE, SRM, DDG, AMM do unscheduled maintenance work.

3. All chapters of AMM are arranged according to Air Transport Association Specification (ATA-100).

4. The SDS gives description of the interfaces, function, configuration and control of the airplane system and subsystems, explains how systems and components operate, shows how systems are constructed, and describes how the interface relates to other systems.

5. Effectivity is the method of defining an airplane configuration and showing configuration differences among airplanes.

Section 4 Aviation Writing

Situations: Yafeng Guo is an aircraft mechanic. He is inspecting the plane and finding some faults. Please help him write down fault description. Some hints of the description words, phrases & terms and the key sentences about fault location, fault description, fault solutions are offered as follows.

Key words, phrases & terms:

1. 为了排故   for troubleshooting
2. 根据   refer to/per/according to
3. 因为   due to/because of
4. 更换   replace
5. 拧紧   tighten
6. 拆下   remove
7. 调整   adjust/regulate
8. 件号   P/N (part number)
9. 放行标准   DDG (dispatch deviation guide)
10. 最低设备清单   MEL (minimum equipment list)
11. 同意放行   dispatch approved/released
12. 因停场时间不足   due to short time
13. 没有备件   lack of parts/no spare parts available/no parts in stock
14. 申请保留   apply for reservation
15. 保留故障   defer defect
16. 保留项目   defer item
17. 保留期限   due time
18. 关闭保留项目   close deferred item

19.	撤消保留项目	rescind deferred item
20.	安装测试	test for installation
21.	操作测试	operational test
22.	系统测试	test for system
23.	地面检查	GND check
24.	试车检查	running-up test
25.	正常	normal / OK
26.	异常	abnormal
27.	与……一致 / 不一致	agree/ disagree with

Key sentences:

1. 过站根据 AMM25-22-00 折下第三排 ABC 座椅，座椅随机放在前货舱。TR:Refer to AMM25-22-00 to remove the passenger seats 3A/3B/3C, and leave in the FWD cargo compartment.

2. 航后依据 AMM23-20-00，更换前服务话筒，并完成功能测试结果正常，撤消 DD/FCXXXXX。AF: Refer to AMM23-20-00 to replace the FWD attendant handset, Accomplish a function test and the result is normal, rescind deferred item DD/FCXXXXX.

3. 航后根据 AMM49-41-11 更换 APU 起动机，启动测试 APU，工作正常。AF: Refer to AMM49-41-11 to replace the APU starter, start APU and do an operation test, the result is normal.

4. 根据 AMMXX-XX-XX 对系统完成测试，结果正常。Refer to AMMXX-XX-XX to do a system function test, the result is normal.

5. 根据 AMM05-51-18 完成冰雹 / 鸟击后的检查，并根据 AMM27-81-00/27-51-00 完成前 / 后缘襟翼的操作测试，结果正常。Refer to AMM05-51-18 to accomplish the Bird/ Hail Strike Conditional Inspection, and refer to AMM27-81-00/27-51-00 to do an operation test of leading edge device and trailing edge flap, the result is normal.

Task: Yafeng Guo has three tasks, the first is to replace the right nose wheel referring to AMM32-45-21, the second is to refer to AMM 12-15-51/301 to check the tire pressure, and ensure the tire pressure is normal, and the last one is to refer to AMMXX-XX-XX to replace the No.2 ATC transceiver.

Module Three Aircraft Structures and Components

Introduction

Airplanes come in many different shapes and sizes, depending on the mission of the aircraft, but all modern airplanes have certain components in common. These are the fuselage, wing, empennage and the control surfaces, landing gear, and powerplant(s). For any airplane to fly, it must be able to lift the weight of the airplane, its fuel, the passengers and the cargo. The wings generate most of the lift to hold the plane in the air. To generate lift, the airplane must be pushed through the air. The engines, which are usually located beneath the wings, provide the thrust to push the airplane forward through the air.

In this module, Boeing 737 is taken for an example to illustrate the main structures and their component of an aircraft.

Fig. M3-1   Boeing 737-700

Lesson 3 Main Structures and Components of an Aircraft

Learning Objectives:

1. Knowledge objectives:

    A. To master the major words, related terms and abbreviations about the main structures and components of an aircraft.

    B. To master the key sentences.

    C. To know the main structures and components of an aircraft.

2. Competence objectives:

    A. To be able to understand frequently-used & complex sentence patterns, acronyms and obtain key information on aviation maintenance quickly.

    B. To be able to talk about the main structures and components of an aircraft in English.

    C. To be able to fill in job cards in English.

3. Quality objectives:

    To be able to self-study with the help of aviation dictionaries, Internet or other resources.

Section 1 Aviation Listening and Speaking

1.1 Aviation listening: listen to the audio and fill in the blanks with the missing words.

听力录音

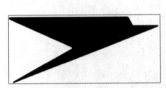  

In 1932, the English designer Lee-Elliott created this very minimalistic 1._____. And by the look of it, it seems to fly really fast, so Mr. Lee Elliott 2._____ to name his design "speed bird".

It immediately became a design classic and was used as the corporate logo for "Imperial Airways" in 1932. The logo was initially 3._____ advertising posters and luggage labels. Later, it was 4._____ the nose section of the company aircraft like the Short S.30 C-Class flying boat.

In 1939 the BOAC (British Overseas Airways Corporation) was 5._____, taking over "Imperial Airways" and smaller British regional aviation companies, and 6._____ the "Speedbird" into their new company logo. 7._____ of the Second World War, most of the BOAC planes were used for military services and were painted in camouflage colors. But the employees insisted to have their speedbird logo visible at the front of the fuselage, as you can see in this picture.

In the 1950s after the World War had ended, the BOAC gave the speedbird greater prominence and placed the logo onto the tail fin, either in colors navy blue on white 8._____ or vice versa. With introduction of proper air traffic control and every airline needing a call sign to identify the operator and aircraft, the BOAC chose the name of their well-known logo, "Speedbird". So the call sign "Speedbird125", the number referring to the flight route and destination, was born. In 1974 the BOAC merged with BEA (British European Airways) and other smaller companies to the world-wide known British Airways.

The speedbird logo remained unaltered but 9._____ to the nose section of the plane to make space for the Union Jack on the tail fin. The new founded British Airways intended to remove the speedbird, but former BOAC employees pressured BA to keep the little bird on the plane.

As in 1984, British Airways prepared for privatization, a new more modern look was needed. The so called "Speedmarque" at the front of the fuselage was designed and bore a minimal resemblance to the original 1930's 10._____. But the call sign remained the same over the years and celebrates it's 66th anniversary this year.

1.2 Aviation Speaking: look at the pictures below and describe them in details.

Clues: cockpit; flight instruments; wing; landing gear; powerplant; empennage.

Section 2 Aviation Reading

Pre-reading questions:

课文朗读录音　　译文

1. How many parts does the fuselage consist of ?

2. Do you know where control surfaces are located? Name all of them please.

Main Structures and Components of an Aircraft
–Taking Boeing 737 as an Example

The Boeing 737 is a short to medium range, single aisle, and narrow body jet airliner with low wings. B737 has semi-monocoque fuselage and fully retractable landing gear. Two powerplants are located under the wings on short struts.

Originally developed as a shorter, lower-cost twin-engine airliner derived from Boeing's 707 and 727, the 737 has nine variants with the -600, -700, -800 and -900 currently in production.

The 737 has been continuously manufactured by Boeing since 1967 with 6,285 aircraft delivered and 2,063 orders yet to be fulfilled as of January 2010. The 737 series is the most-ordered and most-produced jet airliner in history as of April 2009, although the Airbus A320 is the currently most-ordered jet airliner family. There are on average 1,250 737s airborne at any given time, with one departing or landing somewhere every five seconds on average.

Variants of Boeing 737

Boeing 737 Classic

The 737-300 is the same basic design as the 737-200, with a body stretch of 104 inches, a wing tip extension of 14 inches, a horizontal tail extension of 36 inches, a larger dorsal fin and strengthened stabilizer. The 737-400, relative to the 737-300, has body stretch of 120 inches, two additional over-wing exit, tail skid, and strengthened landing gear. The 737-500 uses the 737-300 basic structure with a 94 inches shorter body and a revised forward and aft fairing (wing to body).

微课

Fig.3-1   Boeing737-200

The Boeing 737 Classic refers to the -300/-400/-500 series of the Boeing 737. It is the second generation derivative of the 737, following the original -100/-200 models that began production in 1966. They are short- to medium-range, narrow-body jet airliners. Produced by Boeing Commercial Airplanes from 1984 to 2000, the 737 Classic includes three variants and can seat between 145 and 188 passengers. Improvements over the previous generation of 737 aircraft included CFM International CFM56 high bypass ratio turbofan engines, upgraded avionics, and increased passenger capacity (in the -300/-400 models).

The first model of the Classic series, the 737-300, entered service in 1984. It was followed by a stretched model, the 737-400, which entered service in 1988, followed by shortened 737-500, the smallest variant in the classic series in 1990. In total, 1,988 aircraft were delivered. The Classic series was introduced as the "new generation" of the 737, but following the introduction of the 737 Next Generation in the mid-1990s, it was officially designated as the 737 Classic series.

Lufthansa 737-300, the first 737 Classic model, during climbout after takeoff in 2011

(a) Boeing737-300

(b) Boeing737-400

(c) Boeing737-500

Fig.3-2   737 Classic Series

737 Next Generation

Throughout the 1980s, the 737 Classic series attracted large orders from airlines in the United States and Europe, with its order totals exceeding that of preceding 737 models. By far, the most successful model was the 737-300, with deliveries totaling 1,113 aircraft (the 737-400 and -500 reached 486 and 389 deliveries, respectively). Major operators included US carriers, small national airlines, and charter carriers. However, by the 1990s, with increased competition from the Airbus A320, Boeing began development of a re-winged and updated 737 model which became the 737 Next Generation. Production of the 737 Classic continued alongside that of the Next Generation for a period of time; the first 737-700 was completed in December 1996; the last 737 Classic was completed in February 2000.

(a) Boeing737-600

(b) Boeing737 700

(c) Boeing737-800   (d) Boeing737-900

Fig.3-3   737 NG Series

Main Structures

Fuselage

The fuselage consists of the cockpit, the passenger cabin and the cargo compartments. On cargo planes, the cabin is replaced by a cargo compartment. Fuel is contained in three tanks in the wings and wing center section. In some versions of the planes, an auxiliary tank can be installed in a part of the cargo compartment.

The fuselage is divided into four sections: section 41, section 43, section 46 and section 48. Section 41 contains the radar antenna behind a fiberglass honeycomb fairing, hinged at the top. Aft of the pressure bulkhead, above the floor, are the flight compartment and forward airstair and its door, and the electronic equipment bay. Section 43 contains the passenger cabin and the forward cargo compartment. Section 46 contains the center and aft portion of the passenger cabin, two over-wing escape hatches and aft entry and service doors. The space below the floor includes the wing center section (fuel tank), air-conditioning bays, wheel well, hydraulic bay and aft cargo compartment. This section terminates at the aft pressure bulkhead. Section 48 contains the auxiliary power unit (APU) and horizontal stabilizer truss. Access to this section is through a door on the left side. Aft of which is the APU access door and APU exhaust.

Fig.3-4   Fuselage

The wings

The wings are another main part of the aircraft. The wing provides the principal lifting force of an airplane. Lift is obtained from the dynamic action of the wing with respect to the air. Basic wing structure consists of left, center, and right wing sections. These are built up from the front spar, rear spar, ribs, top and bottom skins and stringers. The wing ribs in the tanks prevent a surge effect in the fuel tank. Various control surfaces are located on the wings including flaps, slats, spoilers, and the ailerons. On most aircraft, leading edge flaps and slats are located on the front of the wings. There are also inboard and outboard flaps on the trailing edge. These control surfaces are hydraulically powered and are only extended or retracted during take-off and landing. Inboard ailerons are for roll control and outboard ailerons are for low airspeed flight. The upper surface of the wing also contains spoilers. They are operated by hydraulic pressure and are used for braking.

Fig.3-5　Wing

The powerplant

The next main part of the plane is the powerplant which consists of the engine and the related accessories. The main engine types are the reciprocating (or piston type), and the reaction, or jet. Usually the engines are attached under the wings, but in the case of the modern aircraft, the two side engines are mounted on struts outboard of the fuselage. The center engine is mounted aft of the fuselage structure below the vertical fin. The inlet for the center engine is above the fuselage, and air intake is through an S-shaped duct.

CFM56

Fig.3-6　The Powerplant

The landing gears

The landing gear, or undercarriage, supports the airplane when it is resting on the ground or in water and during the takeoff and landing. It consists of the nose gear and the main gears. The nose gear is retracted forward into a wheel well in the lower nose compartment. The nose gear is used for maneuvers on the ground. The landing gear is retracted and extended hydraulically. The main gears are attached to the wing rear spar and retract to stow under the main cabin aft of the wing rear spar. The gear may be fixed or retractable. The wheels of most airplanes are attached to shock-absorbing struts that use oil or air to cushion the blow of landing.

Fig.3-7   The Landing Gears

Empennage

Another main part of the aircraft is empennage (also called the tail section). Most aircraft designs consist of a tail cone, fixed surface and movable surfaces. The empennage consists of the vertical stabilizer (sometimes known as the fin), and the horizontal stabilizer. The vertical stabilizer and the horizontal stabilizer are the fixed surfaces. Longitudinal trim is accomplished by movement of the horizontal stabilizer. The movable surfaces are usually a rudder and elevators. The elevators are attached to the trailing edge of this stabilizer and provide additional pitch control. The rudder is used for directional and yaw control.

Fig.3-8   Empennage

Abbreviations and Acronyms

1   Leading Edge (LE)   前缘

| 2 | Trailing Edge (TE) | 后缘 |
| 3 | Auxiliary Power Unit (APU) | 辅助动力装置 |

Words and Phrases

1.	derivative *adj.*	衍生的
2.	jet airliner	喷气式客机
3.	variant *n.*	变体
4.	bypass ratio	涵道比
5.	aisle *n.*	通道
6.	derive *n.*	源于
7.	airborne *adj.*	机载的
8.	stretch *n.*	伸长
9.	semi-monocoque	半硬壳式
10.	fuselage *n.*	机身
11.	cockpit *n.*	驾驶舱
12.	passenger cabin	客舱
13.	cargo compartment	货舱
14.	auxiliary *adj.*	辅助的
15.	install *v.*	安装
16.	radar antenna	雷达天线
17.	fiberglass honeycomb	蜂窝状玻璃纤维
18.	fairing *n.*	整流罩
19.	hinge *v.*	用铰链连接
20.	bulkhead *n.*	隔板，隔离壁
21.	airstair *n.*	登机梯
22.	escape hatch	逃生舱
23.	air-conditioning bay	空调舱
24.	wheel well	轮舱
25.	hydraulic bay	液压舱
26.	terminate *v.*	终止
27.	horizontal stabilizer truss	水平安定面桁架
28.	spar *n.*	翼梁
29.	stringer *n.*	桁条
30.	surge effect	喘振效应
31.	flap *n.*	襟翼
32.	slat *n.*	缝翼
33.	spoiler *n.*	扰流板
34.	aileron *n.*	副翼

35.	vertical stabilizer	垂直安定面
36.	rudder *n.*	方向舵
37.	elevator *n.*	升降舵
38.	inboard *adj.*	内侧的
39.	outboard *adj.*	外侧的
40.	extend *v.*	延伸
41.	retract *v.*	缩回
42.	accessory *n.*	附件
43.	reciprocating *adj.*	往复式
44.	mount *v.*	安装
45.	vertical fin	垂直尾翼
46.	undercarriage *n.*	着陆装置
47.	maneuver *v.*	机动
48.	shock-absorbing strut	减震支柱
49.	cushion *v.*	缓冲
50.	tail *n.*	机尾
51.	longitudinal trim	纵向配平
52.	pitch control	俯仰控制
53.	directional and yaw control	定向和偏航控制

Exercises

答案

2.1 Answer the following questions according to the passage.

1. What does the fuselage consist of ?

2. Where is fuel contained?

3. How many sections is the fuselage divided into?

4. What does empennage consist of ?

5. Where are the engines located?

2.2 Translate the following terms or abbreviations into Chinese.

1. bypass ratio   2. passenger cabin

3. cargo compartment   4. radar antenna

5. fiberglass honeycomb   6. air-conditioning bay

7. wheel well   8. surge effect

9. leading edge   10. longitudinal trim

2.3 Give the equivalent English terms and corresponding Chinese translations according to the pictures.

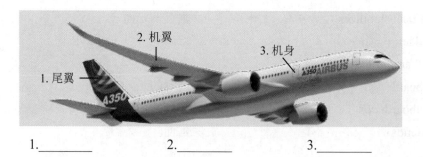

1._____   2._____   3._____

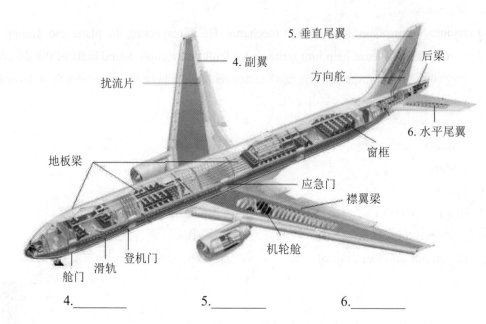

4._____   5._____   6._____

2.4 Find the corresponding translation for the English phrases.

1. horizontal stabilizer truss   A. 减震支柱

2. shock-absorbing strut   B. 半硬壳式

3. elevator   C. 垂直安定面

4. semi-monocoque   D. 升降舵

5. hydraulic bay   E. 后缘

6. vertical stabilizer   F. 水平安定面桁架

7. trailing edge   G. 定向和偏航控制

8. directional and yaw control   H. 俯仰控制

9. vertical fin   I. 液压舱

10. pitch control   J. 垂直尾翼

Section 3 Aviation Translation

Translate the following English sentences into Chinese.

1. The wheels of most airplanes are attached to shock-absorbing struts that use oil or air to cushion the blow of landing.

2. Two powerplants are located under the wings on short struts.

3. In some versions of the planes, an auxiliary tank can be installed in a part of the cargo compartment.

4. They are operated by hydraulic pressure and are used for braking.

5. Usually the engines are attached under the wings, but in the case of the modern aircraft, the two side engines are mounted on struts outboard of the fuselage.

Section 4 Aviation Writing

Situations: Yafeng Guo is an aircraft mechanic. He is inspecting the plane and finding some faults. Please help him write down fault description. Some hints of the description words, phrases & terms and the key sentences about fault location, fault description, fault solutions are offered as follows.

Key words, phrases & terms:

1. 飞机   A/C（aircraft）
2. 机头   nose
3. 机腹   belly
4. 蒙皮   skin
5. 机身   airframe
6. 翼肋   rib
7. 翼梁   spar
8. 机翼   wing
9. 翼尖   wing tip
10. 客舱   cabin / passenger compartment
11. 货舱   cargo compartment
12. 轮舱   wheel well；W/W
13. 驾驶舱   cockpit/flight deck
14. 设备舱   equipment bay
15. 滑窗   sliding window
16. 门窗   door mounted window
17. 过道   aisle
18. 行李架   stowage bin

Key sentences:

1. P46 定检中检查发现右前缘有鸟击现象。

   A bird impact is found at the R/H leading edge at P46 check.

2. 组件灯主警诫再现期间意外亮。

   PACK light comes on unexpectedly during master caution recall.

3. 左发反推不可用。

L/H thrust reverse does not operate.

4. 更换左发外侧反推底部折流门。

Replace L/H ENG outboard T/R bottom block door.

5. 位于座椅扶手的耳机插孔插上耳机后没有声音出来。

There is no sound from the earphone on the seat armrest.

Task: Yafeng Guo finds out three problems in the check. The first, a bird impact is found at the left leading edge at P46 check, the second, the right thrust reverse does not work, and the last, no sound can be heard from the earphone on the right seat armrest.

Lesson 4 Dimensions, Areas and Zones

Learning Objectives:

1. Knowledge objectives:

A. To master the major words, related terms and abbreviations about dimensions, areas and zones.

B. To master the key sentences.

C. To know about dimensions areas and zones.

2. Competence objectives:

A. To be able to understand frequently-used & complex sentence patterns, acronyms and obtain key information about dimensions, areas and zones on aviation maintenance quickly.

B. To be able to talk about aviation or aircraft in English.

C. To be able to fill in job cards in English.

3. Quality objectives:

To be able to self-study with the help of aviation dictionaries, Internet or other resources.

Section 1 Aviation Listening and Speaking

1.1 Aviation Listening: listen to the audio and fill in the blanks with the missing words.

听力录音

Do airplanes have windscreen wipers such as cars? Put yourself into your 1._____ seat and imagine that we are slowly taxing towards the runway and it's raining heavily. A passenger jet taxi's (straight) with a speed at about 20-35 mph (30-60 km/h), which isn't 2._____ enough for the airstream to clear the windshields from raindrops. The visibility reduces dramatically which can be dangerous on airport aprons. So with one little hand movement, like in your car, every pilot can 3._____ on his windscreen wiper. There is one wiper on each side which can be turned on separately and has two speed settings, slow and fast. During a rainy take-off roll we keep the wipers switched on, cause as you gain speed a lot of the water from the nose cone runs over the windshields and the visibility is reduced even more, so you have the option to increase the wiper speed.

4._____ as you go along the motorway at 60 mph, you may use wiper speed, but if you accelerate to 100mph, at the same amount of rain, you're going to have to adjust your wiper speed.

Okay let's imagine we're in flight.

During enroute cruise the wipers are inoperative. They are aerodynamically and mechanically limited to the 5._____ of the aircraft so that the switches are inhibited. The limit varies between aircraft manufactures. The limiting speed for example on an Airbus A320 is 230 kts (260 mph). On final approach (approx 20 miles from the runway) you 6._____ at speeds of 260 mph, which is fast enough to clear the windshields from rain. But three miles prior to 7._____ you will have reduced the speed down to 140 kts and in heavy rain the wipers are absolutely essential to get clearly visibility onto the runway.

In the event that one wiper is inoperative, may only the pilot with the 8._____ wiper fly the aircraft. In the event that both wipers are inoperative, which is very unlikely, the aircraft is limited such as it may only 9._____ and land in visual metrological conditions and it's 10._____ to land within certain approach minimas (height and visibility) is reduced and limited.

1.2 Aviation Speaking: look at the pictures below and describe them in details.

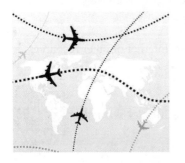
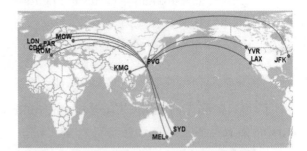

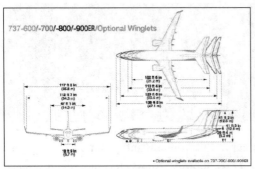

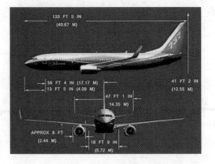

Clues: range; design range; capacity; structure dimensions.

Pre-reading questions:

1. How many engines are there on a Boeing 737-800 airplane?

2. How many passengers can the plane hold at most?

课文朗读录音　　译文

Dimensions, Areas and Zones

The 737-800 is a two-engine airplane for short to medium range flights with a capacity of up to 189 passengers. This aircraft has many new features such as the payload, service ceiling and range. It has a design range of 2,900 nautical miles.

Fig. 4-1 Boeing 737-800

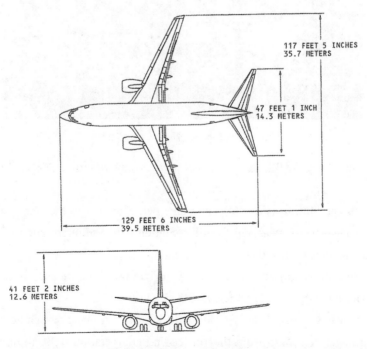

Fig. 4-2   Structure Dimensions

The following material contains information on principal dimensions for the wing, ailerons, flaps, horizontal stabilizer surfaces, vertical stabilizer surfaces and fuselage. It also includes information on areas for the wing and tail surfaces. It supplies station diagrams for the wing, vertical tail surfaces and engine nacelle.

Dimension

The airplane is divided into reference planes designated as stations, waterlines and buttock lines measured in inches. This provides a way of quickly identifying the location of components, the center of gravity and the distribution of the weight.

Body Station (BSTA): The plane perpendicular and measured parallel to the body centerline from a point 130 inches forward of the nose.

Body Buttock Line (BBL): The plane measured perpendicular to the body vertical centerline plane, BBL 0.00.

Body Waterline (BWL): The plane measured perpendicular to a horizontal plane located 148.5 inches below the body, BWL 0.00.

微课

Fig. 4-3　Boeing 737-800 in a Hangar

Body Reference Plane (BRP): The horizontal plane, Body Waterline 208.1, at the top surface of the floor beams.

Vertical Stabilizer Station (Fin Sta): The plane perpendicular to the centerline of the vertical stabilizer rear spar. Distance is measured from Fin Station 0.00, the intersection of the leading edge line extension and the fin waterline 0.00.

Vertical Stabilizer Waterline (Fin WL):A horizontal plane measured parallel to a Body Waterline. Fin Waterline 0.00 is Body Waterline 300.50.

Vertical Stabilizer Leading Edge Station (Fin LE Sta): A plane perpendicular to the vertical stabilizer leading edge, measured from the fin leading edge station 0.00, the intersection of the leading edge line extension and the vertical stabilizer waterline 0.00.

Rudder Station(Rud Sta):A plane perpendicular to the rudder hinge centerline, measured from

Rudder Station 0.00, the intersection of the rudder hinge centerline and vertical stabilizer waterline 0.00.

Horizontal Stabilizer Station(Horizontal Stab Sta): A plane perpendicular to stabilizer chord plane and plane of stabilizer rear spar, measured from stabilizer station 0.00, the intersection of the leading edge and stabilizer buttock line 0.00.

Stabilizer Chord Plane (SCP): A plane through the trailing and leading edges of the stabilizer airfoil.

Next is the dimensions and areas of 737-800.

Fuselage dimensions

Dimensions give locations on the fuselage in inches. Use the following dimensions to find components on the fuselage: Body station line; Body buttock line and Water line.

The body station line (STA) is a horizontal dimension which starts at station line zero. Measure the body station line for a vertical reference plane which is forward of the airplane.

The body buttock line (BL) is a lateral dimension. Measure the buttock line to the left or right of the airplane center line.

The water line (WL) is a height dimension. Measure the water line for a horizontal reference plane below the airplane.

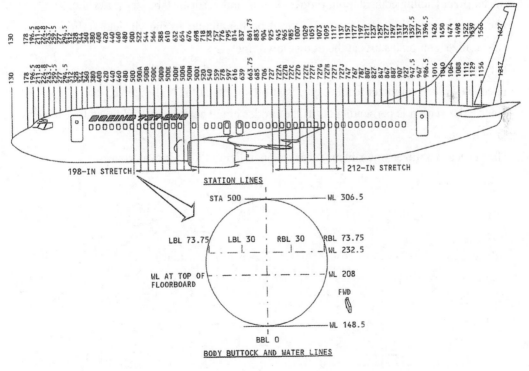

Fig. 4-4   Fuselage Dimensions

Wing reference dimensions

The wing has two reference dimensions, giving wing location in inches.

Measure each location from buttock line 0. They are wing station and wing buttock line. Measure the wing buttock line parallel to the buttock line.

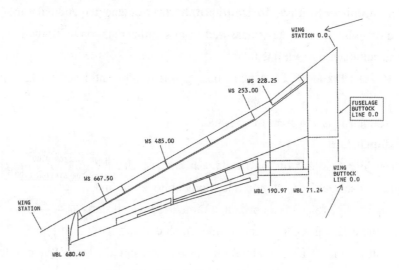

Fig. 4-5　Wing Reference Dimensions

Vertical stabilizer reference dimensions

The vertical stabilizer has four reference dimensions (in inches): vertical stabilizer station, vertical stabilizer leading edge station, rudder station, and vertical stabilizer waterline.

The vertical stabilizer station is measured perpendicular to the vertical stabilizer rear spar. Vertical stabilizer station 0 starts at the body crown line.

The vertical stabilizer leading edge station is measured perpendicular to the vertical stabilizer leading edge. Vertical stabilizer leading edge station 0 starts at the body crown line.

The rudder station is perpendicular to the rudder hinge centerline. Rudder station 0 starts at the body crown line.

The vertical stabilizer waterline is measured parallel to the body waterline.

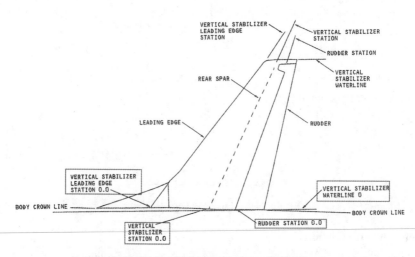

Fig. 4-6　Vertical Stabilizer Reference Dimensions

Horizontal stabilizer reference dimensions

The horizontal stabilizer has three reference dimensions (in inches): stabilizer station, stabilizer leading edge station and elevator station. Each location is measured from buttock line 0.

Stabilizer stations are measured perpendicular to the horizontal stabilizer rear spar.

Stabilizer leading edge stations is measured perpendicular to the horizontal stabilizer leading edge.

Elevator stations are measured perpendicular to the elevator hinge centerline.

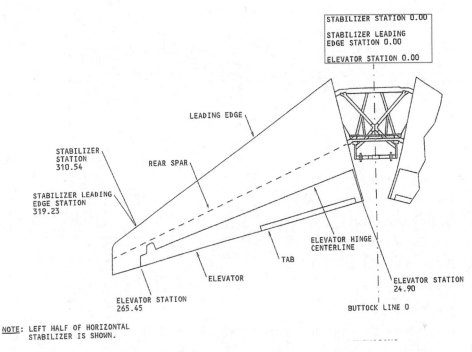

Fig. 4-7   Horizontal Stabilizer Reference Dimensions

Zone Diagram

Eight major zones for the airplane help you find the airplane components and parts, they are:

1) 100 – lower half of the fuselage

2) 200 – upper half of the fuselage

3) 300 – empennage

4) 400 – powerplant and nacelle struts

5) 500 – left wing

6) 600 – right wing

7) 700 – landing gear and its doors

8) 800 – doors

The major zones are divided into sub-zones and the sub-zones are divided into zones.

Information on the section of numbers, zone numbers, and access panel numbers is in the maintenance planning document (MPD).

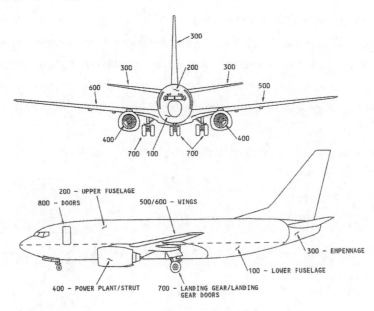

Fig. 4-8　Zone Diagram

Abbreviations and Acronyms

1.	Body Reference Plane (BRP）	机身基准面
2.	Buttock Line (BL)	纵剖线
3.	Left Buttock Line (LBL)	左机身纵剖线
4.	Right Buttock Line (RBL)	右机身纵剖线
5.	Station (STA)	站位
6.	Waterline (WL)	水线
7.	Maintenance Planning Data Document (MPD)	维修计划数据文献

Words and Phrases

1.	medium *adj.*	中间的；中等的
2.	capacity *n.*	容量
3.	payload *n.*	商载
4.	service ceiling	实用升限
5.	nautical mile	海里
6.	body *n.*	机身
7.	fuselage *n.*	机身
8.	station line	站位线
9.	buttock line	纵剖线
10.	water line	水线
11.	reference plane	基准面
12.	lateral *adj.*	横向的
13.	parallel to	与……平行
14.	vertical stabilizer	垂直安定面

15.	rudder *n.*	方向舵
16.	perpendicular to	与……垂直
17.	spar *n.*	翼梁；横梁
18.	crown line	顶线
19.	hinge *n.*	铰链
20.	horizontal stabilizer	水平安定面
21.	elevator *n.*	升降舵
22.	empennage *n.*	机尾
23.	nacelle *n.*	吊舱

答案

Exercises

2.1 Answer the following questions according to the passage.

1. How many passengers can Boeing 737-800 airplane hold at most?
2. What is STA?
3. How many reference dimensions does the vertical stabilizer have?
4. How do we measure stabilizer stations?
5. How many major zones does an airplane have and what's the usage of major zones?

2.2 Find the corresponding translation for the English phrases.

1.	station line	A. 吊舱
2.	buttock line	B. 机身顶线
3.	water line	C. 纵剖线
4.	rudder	D. 升降舵
5.	vertical stabilizer	E. 垂直安定面
6.	horizontal stabilizer	F. 水位线
7.	body crown line	G. 水平安定面
8.	hinge	H. 站位线
9.	elevator	I. 方向舵
10.	nacelle	J. 铰链

2.3. Give the equivalent English terms and corresponding Chinese translations according to the pictures.

1._____

2._____

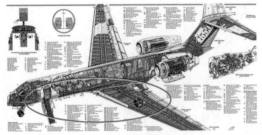

3._____

4. _____

5. _____

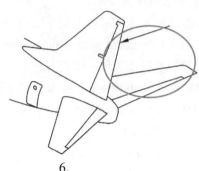

6. _____

Section 3 Aviation Translation

Translate the following English sentences into Chinese.

1. These new features increase the airplane payload, service ceiling, and range.
2. They are vertical stabilizer station, vertical stabilizer leading edge station, rudder station, and vertical stabilizer waterline.
3. The body buttock line (BL) is a lateral dimension.
4. The wing has two reference dimensions, giving wing location in inches.
5. The airplane has eight major zones to help you find the airplane components and parts.

Section 4 Aviation Writing

Situation: Yafeng Guo is an aircraft mechanic. He is inspecting the plane and finding some faults. Please help him write down fault description for him. Some hints of the description words, phrases & terms and the key sentences about fault location, fault description, fault solutions are offered as follows.

Key words, phrases & terms:

1. 检查：check
2. 发现：find
3. 航后检查：AF check
4. 过站检查：TR check
5. 故障：fail
6. 超标：out of limits

7. 磨损：worn

8. 漏油：leakage

Key sentences of finding faults:

······检查发现 / 报告故障······

1. 检查发现 2 号 ATC 故障。

   Check: ATC system No.2 ATC fails.

2. 检查发现 2 号 ATC 故障。

   Check: ATC system No.2 ATC fails.

3. 检查发现左外主起落架刹车组件磨损超标。

   Check: L outboard wear indicator pin of MLG brake assy is out of limits.

4. 过站航后检查发现：检查发现左外刹车组件磨损超标。

   TR check: L outboard wear indicator pin of MLG brake assy is out of limits.

5. 检查发现左主起落架一液压管漏油超标。

   Check: the L MLG hydraulic tube leakage is out of limits.

Task: Yafeng Guo find out three problems in AF check, the first, main landing gear brake assembly is out of limits, the second, the left after position light is not on, and the last, the "Occupied/None" placard in forward lavatory is loosening. Please finish the fault description.

Lesson 5 Panels in the Flight Compartment

Learning Objectives:

1. Knowledge objectives:

A. To master the major words, related terms and abbreviations about the main panels in the flight compartment.

B. To master the key sentences.

C. To know the panels in flight compartment.

2. Competence objectives:

A. To be able to understand frequently-used & complex sentence patterns, acronyms and obtain key information on the main panels in the flight compartment quickly.

B. To be able to talk about aviation or aircraft in English.

C. To be able to fill in job cards in English.

3. Quality objectives:

To be able to self-study with the help of aviation dictionaries, Internet or other resources.

Section 1 Aviation Listening and Speaking

1.1 Aviation Listening: listen to the audio and fill in the blanks with the missing words.

听力录音

Today's topic will all be about the correct 1._____ seating 2._____ in the Airbus, using the Eye-Position-Indicator. Then I'll 3._____ you how to adjust the seat in its 4._____ and horizontal axis. Many of you might not have known that the Airbus cockpit seat can be adjusted 5._____ and manually.

The correct seating position is 6._____ for your visual segment regarding your view onto the PFD-Primary Flight 7._____ and the ND- 8._____ Display, and the other 50% of your peripheral vision will concentrate on the 9._____.

Then I'll show you how to adjust and set the armrest and how your forearm should be placed in accordance with the side stick.

We'll also look at the 10._____ pedals, how they are adjusted and importance of reaching the upper part of the pedals to apply brake pressure!

1.2 Aviation Speaking: look at the pictures below and describe them in details.

Clues: panels; lights and indicators; abbreviations; communication; navigation.

Section 2 Aviation Reading

Pre-reading questions:

1. Have you ever seen a flight compartment in a picture or in a movie?

2. What does a flight compartment contain?

课文朗读录音

译文

Fig 5-1 shows different panels in the flight compartment.

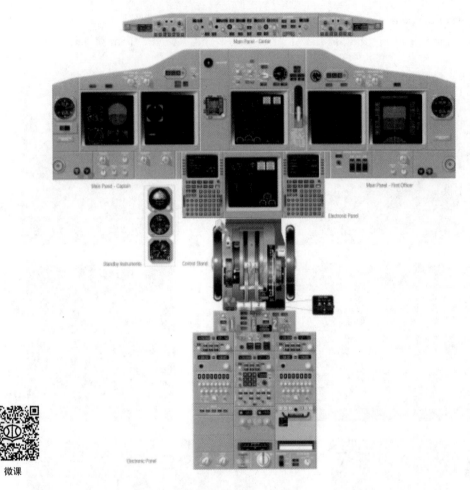

Fig. 5-1   Different Panels in the Flight Compartment

Introduction

These are the major panels in the flight compartment:

- ➤ P1 captain instrument panel
- ➤ P2 center instrument panel
- ➤ P3 first officer instrument panel
- ➤ P5 forward overhead panel
- ➤ P5 aft overhead panel
- ➤ P7 glareshield panel
- ➤ P8 aft electronic panel
- ➤ P9 forward electronic panel
- ➤ Control stand

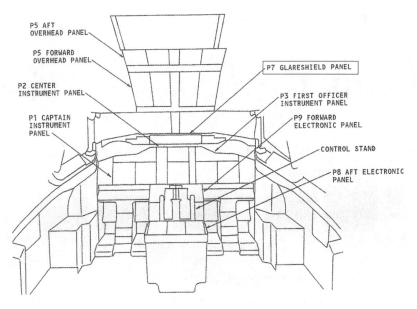

Fig. 5-2 Locations of Different Panels

Main Instrument Panels

P1&P3 consist of the main instrument panels. The controls and displays on this panel include these components:

- Clock (2)
- Display unit (4)
- Display select panel (2)
- Master dim and test switch
- Brake pressure indicator
- Autoflight status annunciator
- GPWS control panel
- Lighting control (2)
- Conditioned air outlet control (2)

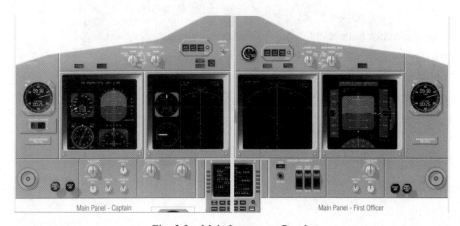

Fig. 5-3 Main Instrument Panels

P7 Glareshield Panel

Fig. 5-4　P7 Glareshield Panel

The controls and displays on the P7 glareshield include these components:

➢　Master caution annunciator (2)

➢　System caution annunciator (2)

➢　Mode Control Panel (MCP)

➢　EFIS control panel (2)

➢　Fire warning light (2)

P2 Center Instrument Panel & P9 Forward Electronics Panel

The controls and displays on the P2 center instrument panel include these components:

➢　Yaw damper indicator

➢　Standby instruments

➢　Engine display control panel

➢　Antiskid and autobrake switches and lights

➢　Flap position indicator

➢　Landing gear lever and position indicator

➢　Upper center display unit

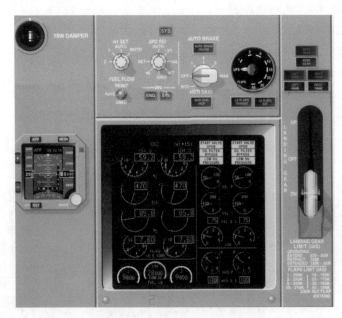

Fig. 5-5　P2 Center Instrument Panel

The controls and displays on P9 forward electronics panel include these components:

> Lower center display unit

> Multi-function control display unit (MCDU) (2)

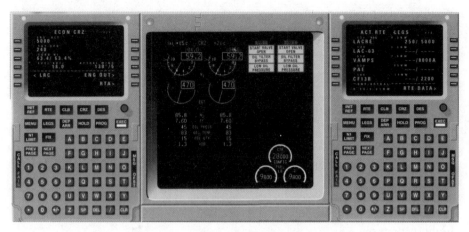

Fig. 5-6　P9 Forward Electronics Panel

Control Stand

The controls and indications on the control stand include these components:

> Forward thrust lever

> Reverse thrust lever

> Speed brake handle

> Horizontal stabilizer trim wheel and indicator

> Parking brake lever and light

> Flap lever

> Stabilizer trim cutout switch

> Start lever

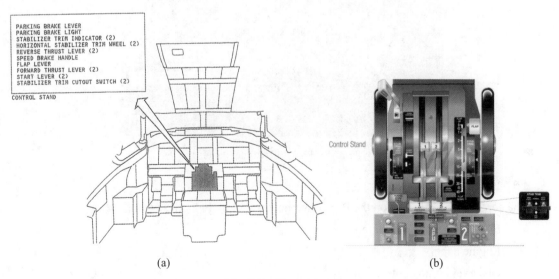

(a)　　　　　　　　　　　(b)

Fig. 5-7　Control Stand

P8 Aft Electronics Panel

The P8 aft electronics panel has these components:

- ➢ Weather radar control panel
- ➢ Cargo fire control panel
- ➢ Aileron/rudder trim panel
- ➢ ATC/TCAS control panel
- ➢ Audio control panel
- ➢ Overheat/fire protection panel
- ➢ Navigation control panel
- ➢ Radio communication panel
- ➢ ADF control panel
- ➢ ACMS printer
- ➢ Lighting control

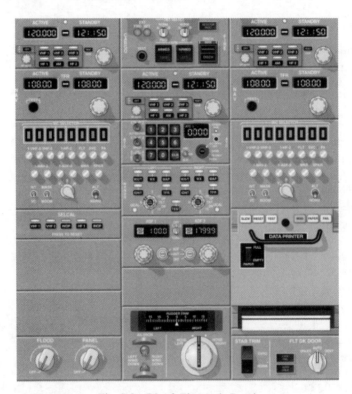

Fig. 5-8   P8 Aft Electronic Panel

P5 Aft Overhead Panel

The controls and displays on the P5 aft overhead panel include these components:

- ➢ Inertial system display unit
- ➢ Engine panel
- ➢ Observer audio control panel
- ➢ Oxygen panel

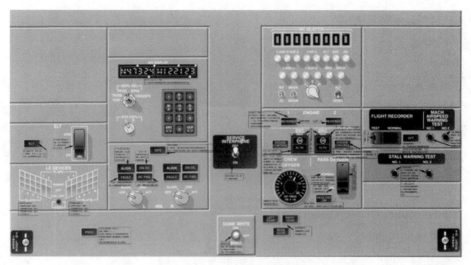

Fig. 5-9   P5 Aft Overhead Panel

- ➢ Landing gear indicator lights
- ➢ White dome light switch
- ➢ Service interphone switch
- ➢ IRS mode select unit
- ➢ Flight recorder and mach airspeed warning module
- ➢ Stall warning test module
- ➢ Proximity switch electronics unit light
- ➢ Leading edge devices annunciator panel

P5 Forward Overhead Panel

The controls and displays on the P5 forward overhead panel include these components:

- ➢ APU control switch
- ➢ APU indicator panel
- ➢ Fuel control panel
- ➢ Ground power and bus switching panel
- ➢ Equipment cooling panel
- ➢ Generator drive and standby power panel
- ➢ AC and DC meter panel
- ➢ Flight control panel
- ➢ Air-conditioning/bleed air control panel
- ➢ Hydraulic control panel
- ➢ Cabin altitude panel
- ➢ Cabin pressure control panel
- ➢ Cockpit voice recorder panel
- ➢ Engine start panel
- ➢ Passenger signs panel

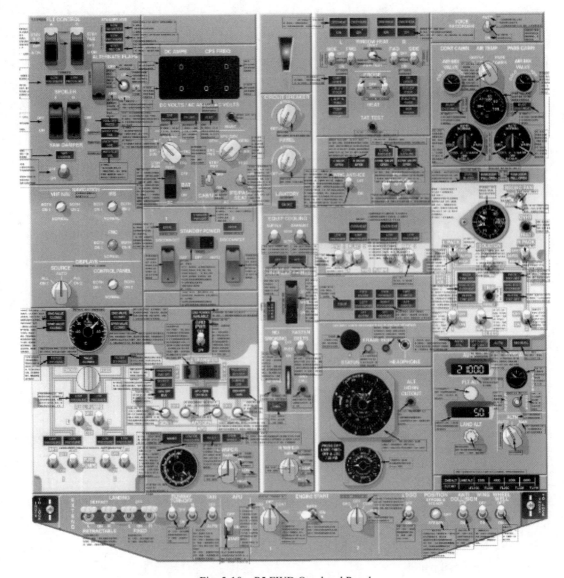

Fig. 5-10　P5 FWD Overhead Panel

- ➢ Source select panel
- ➢ Window/pitot heat module
- ➢ Instrument switching and VHF Nav and IRS panel
- ➢ Door warning panel
- ➢ Anti-ice panel
- ➢ Temperature control panel
- ➢ Light switches

Aft Flight Compartment

The main circuit breaker panels are behind the first officer and captain seats. The P6 and P18 have the component load circuit breakers. Circuit breakers are organized by airplane systems. The P61 panel has the data loader controls.

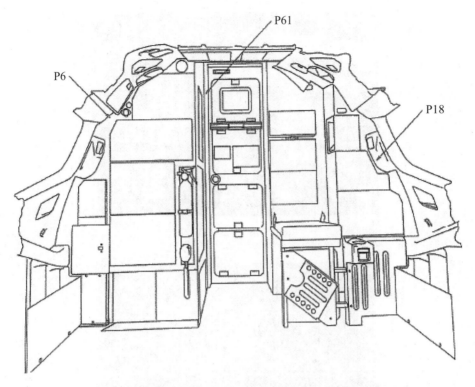

Fig. 5-11　Aft Flight Compartment

Fig. 5-12　P18 Panel

Fig. 5-13　P18-3 Panel

Fig. 5-14　P61 Panel

Fig. 5-15　P6-11 Panel

Fig. 5-16 P21 Panel

Abbreviations and Acronyms

1. Air Traffic Control (ATC)   空中交通管制
2. Automatic Direction Finder (ADF)   自动定向仪
3. Aircraft Condition Monitoring System (ACMS)   飞行状态监控系统
4. Electronic Flight Instrument System (EFIS)   电子飞行仪表系统
5. Ground Proximity Warning System (GPWS)   近地警告系统
6. Inertial Reference System (IRS)   惯性基准系统
7. Traffic Alert and Collision Avoidance System (TCAS) 交通警告和防撞系统
8. Multi-function Control Display Unit (MCDU)   多功能控制显示器组件

Words and Phrases

1. glareshield *n.*   遮光板
2. captain *n.*   机长
3. instrument *n.*   仪表
4. panel *n.*   面板
5. first officer   副机长；副驾驶
6. control stand   控制台
7. dim *adj.*   暗的，暗亮的
8. annunciator *n.*   信号牌
9. yaw damper   偏航阻尼器
10. thrust lever   推力杆；油门杆
11. trim wheel   配平轮
12. parking brake   停留刹车
13. inertial *n.*   惯性的
14. dome light   舱顶灯
15. mach *n.*   马赫
16. stall *n.*   失速
17. bleed air   引气
18. altitude *n.*   海拔；高度

19. autoflight *n.*　　　　自动飞行
20. antiskid *adj.*　　　　防滑的
21. pitot *n.*　　　　空速管
22. circuit *n.*　　　　电路
23. outlet *n.*　　　　出口

Exercises

答案

2.1 Answer the following questions according to the passage.

1. What do the main instrument panels consist of?
2. Where is the parking brake lever?
3. Where is the overheat/fire protection panel?
4. Where are the main circuit breaker panels?
5. Does the P61 panel have the data loader controls?

2.2 Translate the following terms or abbreviations into Chinese.

1. captain instrument panel
2. ACMS
3. GPWS
4. first officer
5. yaw damper
6. thrust lever
7. ATC
8. control stand
9. glareshield
10. parking brake

2.3 Give the equivalent English terms and corresponding Chinese translations according to the pictures.

1._____

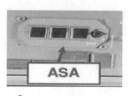

2._____

3._____

4._____

5._____

6._____

Section 3 Aviation translation

Translate the following English sentences into Chinese.

1. The controls and displays on the main instrument panels (P1&P3) include these components.

2. The main circuit breaker panels are behind the first officer and captain seats.

3. Circuit breakers are organized by airplane systems.

4. The P6 panel has the data loader controls.

5. The P6 and P18 have the component load circuit breakers.

Section 4 Aviation Writing

Situations: Yafeng Guo is an aircraft mechanic. He is inspecting the plane and finding some faults. Please help him write down fault description. Some hints of the description words, phrases & terms and the key sentences about fault location, fault description, fault solutions are offered as follows.

Key words, phrases & terms:

1. 关闭 close / shut
2. 连接 connect
3. 断开 disconnect
4. 松开 release
5. 减少 decrease
6. 增加 increase
7. 拆除 remove
8. 确保 ensure / make sure
9. 设置 set
10. 安装 install / mount
11. 松开 loosen
12. 拧紧 tighten
13. 开始，起动 start
14. 定位 locate / position

15. 供应　supply / provide

16. 打开，断开　open

Key sentences:

1. 自动驾驶仪接通时操纵杆（盘）移动。

   Control columns (wheels) move when the autopilot engages.

2. 偏航阻尼器自动断开。

   Yaw damper disengaged automatically.

3. A/P 红色警告信号牌灯恒亮。

   A/P red warning annunciator light on steady.

4. 飞机准备时偏航阻尼电门接不通。

   The YAW DAMPER switch could not be set on when preparing for taking off.

5. 在自动驾驶多通道进近期间。Found during autopilot multi-channel approach.

Task: Yafeng Guo finds out three problems in the check. The first, disengage switch doesn't disengage when it is pushed, the second, the right thrust reverse does not operate, and the last, there is no sound from the earphone on the right seat armrest. Please finish the writing tasks for him.

Lesson 6 Different Doors of an Aeroplane

Learning Objectives:

1. Knowledge objectives:

A. To master the major words, related terms and abbreviations about different doors.

B. To master the key sentences.

C. To know the door types and process of opening and closing entry doors.

2. Competence objectives:

A. To be able to understand frequently-used & complex sentence patterns, acronyms and obtain key information on aviation maintenance quickly.

B. To be able to talk about aviation or aircraft in English.

C. To be able to fill in job cards in English.

3. Quality objectives:

To be able to self-study with the help of aviation dictionaries, Internet or other resources.

Section 1 Aviation Listening and Speaking

听力录音

1.1 Aviation Listening: listen to the audio and fill in the blanks with the missing words.

After 1._____ and during climb, the outside air pressure is in a constant fall, whilst the air pressure within the 2._____ is in a constant rise created by the aircraft's bleed and air conditioning system, otherwise you would have trouble breathing during the flight. Due to the big

difference between the ambient pressure in the cabin and the low pressure outside, there is a lot of 3._____ stress on the cabin hull and especially on the windows.

Now if you take a closer look at the cabin windows you'll see they are made up out of 4._____ separate window panes. One outer and 5._____ pane which share the same window 5._____ and one inner pane.

Now if you look at your windows installed in your home you'll see they are also made up out of at least two panes of glass, with a transparent 7._____ in between them, acting as a temperature barrier which stops condensation on the inner side of the window.

Same in the airplane but without the special gas in between. Because the air pressure difference and temperature variations are far much 8._____ than in your home, engineers had to come up with an idea to prevent the windows from fogging and 9._____ up. So they've decided to leave an air gap in between the two panes. But air has the physical effect of expanding and shrinking either due to air pressure or temperature changes, which could 10._____ the windows.

1.2 Aviation Speaking: look at the pictures below and describe them in details.

Clues: types of doors; control handle; door window; escape slide.

课文朗读录音　　译文

Section 2　Aviation Reading

Pre-reading questions:

1. How many types of doors are there on the airplane?

2. Do you know where emergency exit doors are located?

Different Doors of an Aeroplane

The doors are movable units which give access to the airplane compartments. These are different types of doors:

> ➢ Forward and aft entry doors
>
> ➢ Forward and aft galley service doors
>
> ➢ Emergency exit doors
>
> ➢ Cargo doors
>
> ➢ Miscellaneous access doors

A door warning system shows the crew that pressure bearing doors are closed and properly latched before flight.

Pressure doors have silicon rubber seals. The seals have such functions as:

> ➢ Seal air and light leaks
>
> ➢ Act as acoustic and thermal barriers
>
> ➢ Supply aerodynamic smoothness.

Location

The doors on the left side of the airplane are entry doors. Those on the right side are galley service doors. The emergency exit doors are above the wings on both sides of the airplane. The cargo doors are on the right side. The miscellaneous access doors are located near the systems they serve.

Fig. 6-1　Entry Door

You can open and close entry doors, galley service doors and cargo doors if the wind is up to 40 knots with no damage to the structure. You can let the doors stay latched open if the wind is up to 65 knots with no damage. A protective cover should be put over the door frame to prevent bad weather damage to the airplane if a door keeps open for a long time. A safety strap must be attached in the doorway when entry and galley service doors are open and not used.

Fig. 6-2　Emergency Exit Door

Fig. 6-3　Cargo Door

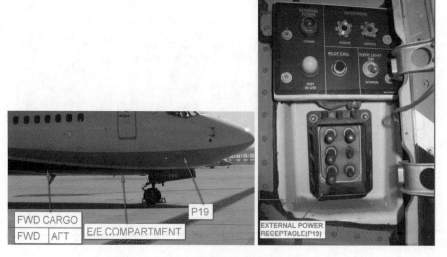

Fig. 6-4　External Power Receptacle Door

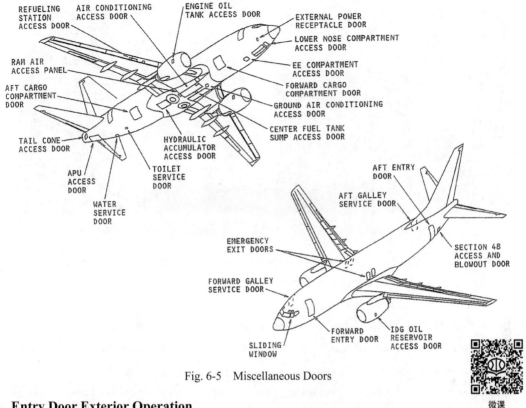

Fig. 6-5　Miscellaneous Doors

Entry Door Exterior Operation

There is an escape slide on the inside lower part of the entry door. If the slide girt bar is armed, the slide will deploy automatically as the door opens. The cabin crew should put the warning pennant across the door window when the slide is armed. Examine if there is a door slide warning pennant (orange) in the door window before operation.

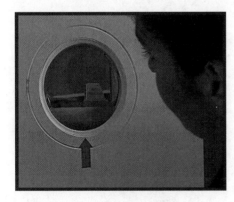

Fig. 6-6　Observation Window

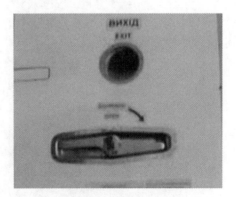

Fig. 6-7　Control Handle

Opening operation

Before opening operation, check the door slide warning pennant in the door window to make sure that the escape slide is not armed.

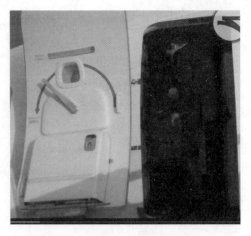

Fig. 6-8　Escape Slide　　　　Fig. 6-9　Escape Slide Container

Use a control handle in the recess to open and close an entry door. First pull the exterior door control handle to engage the door drive mechanisms. Then turn the handle 180 degree in the clockwise direction. The door moves to the cocked open position. At this position, any more effort on the door control handle will not produce more door motion because it has gone through its full motion. Next release the control handle and let it return to its recess by spring force. The door will swing out of the door frame.

As you pull the door open, the door turns. To complete the door opening operation, you should hold the aft edge and pull it open. This moves the door parallel to the airplane fuselage when the door is fully open. A hinge lock mechanism holds the door in the open position while the door is fully open.

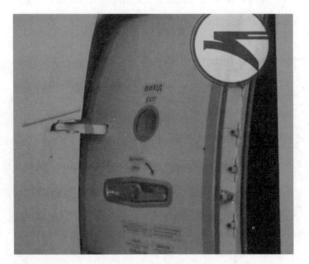

Fig. 6-10　Entry Door Swinging out of the Frame

NOTE: MAKE SURE THAT THE AREA OUTSIDE OF THE DOOR IS CLEAR. AS THE DOOR OPENS, WINDS MAY PUSH IT. THIS CAN PUSH THE OPERATOR OFF BALANCE. SO THE OPERATOR SHOULD KEEP A STRONG AND SECURE STANCE.

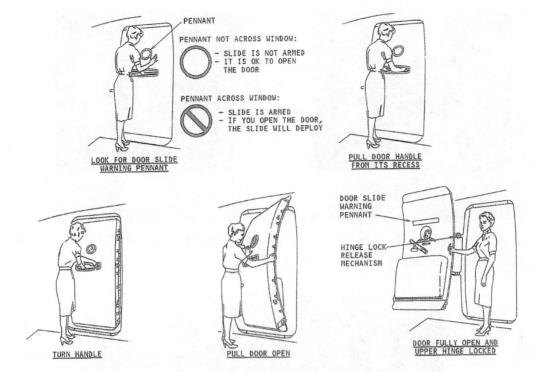

PENNANT

PENNANT NOT ACROSS WINDOW:
- SLIDE IS NOT ARMED
- IT IS OK TO OPEN THE DOOR

PENNANT ACROSS WINDOW:
- SLIDE IS ARMED
- IF YOU OPEN THE DOOR, THE SLIDE WILL DEPLOY

LOOK FOR DOOR SLIDE WARNING PENNANT

PULL DOOR HANDLE FROM ITS RECESS

TURN HANDLE

PULL DOOR OPEN

DOOR SLIDE WARNING PENNANT

HINGE LOCK RELEASE MECHANISM

DOOR FULLY OPEN AND UPPER HINGE LOCKED

Fig. 6-11 Entry Door Exterior Operation

Closing operation

Examine the escape slide first before closing the door. The slide strap and bar must be properly stowed for an incorrectly folded strap or improperly stowed bar that interfere with the door threshold, preventing the door from sealing and latching with damage to components.

Then release the hinge lock with the yellow latch release mechanism. This will swing the door back into the door frame. Pull the door to the cocked position. Next pull the exterior control handle, and turn it slightly until it engages the door drive mechanisms. Turn the door control handle counterclockwise 180 degrees. Then release it and allow it to the recess by spring force.

Training information point

The force to open and close the door on the control handle is not large. If a large force is needed, there must be a fault with the door or the procedure. Check if the door-to-frame area is clear or if an incorrectly stowed escape slide girt strap is caught between the door and the frame.

If the airplane is pressurized, a properly rigged door will not unlatch. This is because the door gates must open against cabin pressure. Pressure on the door gates prevents this.

Entry Door Interior Operation

Before opening operation from the interior, check if there is a door slide warning pennant in the door window and remove it if there is. Then unlatch the door with the interior control handle and turn the handle in the OPEN direction. Push the door through the door frame until it is fully open. Use the assist handles for this operation. A hinge lock mechanism in the upper hinge keeps the door in the fully open position.

Fig. 6-12　The OPEN Direction

Fig. 6-13　The Door through the Door Frame

To close the door, first operate the yellow latch release mechanism to unlock the hinge. This will swing the door back into the door frame. Pull the door to the cocked position and turn the control handle in the close direction.

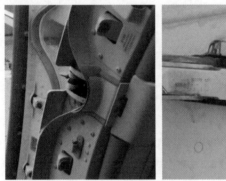
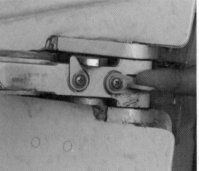

Fig. 6-14　The Hinge Lock Mechanism

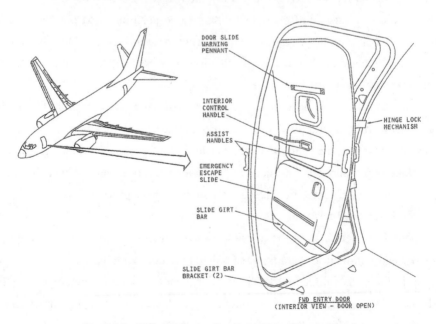

Fig. 6-15　Different Parts of an Entry Door (InteriorView)

Abbreviations and Acronyms

1. Forward (FWD)   前面的
2. After (AFT)   后面的
3. Integrated Drive Generator (IDG)   整体驱动发电机

Words and Phrases

1. entry door   登机门
2. galley service door   厨房勤务门
3. emergency exit door   紧急出口门
4. cargo door   货舱门
5. miscellaneous *adj.*   各种各样的
6. access door   接近门
7. flight control   飞行操纵
8. crew *n.*   机组人员，（全体）工作人员
9. pressure bearing door   承压门
10. latched *adj.*   锁住的；上了锁的
11. silicon rubber seal   硅胶密封垫
12. knot *n.*   节（风速单位）
13. protective cover   防护罩
14. safety strap   安全带
15. doorway   门口
16. escape slide   逃生滑梯
17. girt bar   束缚杆
18. armed *adj.*   预位的
19. door slide warning pennant   门滑梯警示牌
20. clean *adj.*   无障碍的
21. recess *n.*   凹槽
22. engage *v.*   （使机器）接合；啮合
23. drive mechanism   驱动装置
24. cocked *adj.*   准备的，向上翘起的
25. effort *n.*   努力
26. motion *n.*   位移
27. release *v.*   松开
28. spring force   弹簧弹力
29. balance *n.*   平衡
30. secure stance   安全的站姿
31. girt strap   束缚带
32. threshold *n.*   门槛
33. counterclockwise *adj.*   逆时针的

34. procedure *n.*		操作程序
35. stow *v.*		收藏
36. pressurize *v.*		增压；给…加压
37. rig *v.*		给（飞机、船）校准
38. hinge lock mechanism		铰链锁定装置
39. assist handle		辅助手柄

Exercises

2.1 Answer the following questions according to the passage.

1. How many types of doors are there on the airplane?

2. Where is emergency exit door?

3. What should the operator do to prevent bad weather damage to the airplane if a door keeps opening for a long time?

4. What should the operator do to keep balance as he/she pulls the door open?

5. What should the operator do before opening operation from the interior?

2.2 Tell True (T) or False (F) according to the passage.

(   ) 1. These are six types of doors.

(   ) 2. The cargo doors are on the left side.

(   ) 3. To complete the door opening operation, you should hold the forward edge of the door and pull it open.

(   ) 4. If the airplane is pressurized, a properly rigged door will not unlatch.

(   ) 5. To open and close the door, the force on the control handle is large.

2.3 Translate the following terms or abbreviations into Chinese.

1. galley   2. control handle

3. recess   4. hinge mechanism

5. thermal   6. silicon rubber seal

7. aerodynamic   8. girt bar

9. counterclockwise   10. escape slide

2.4 Give the equivalent English terms and corresponding Chinese translations according to the pictures.

1.＿＿＿＿＿＿＿

2.＿＿＿＿＿＿＿

3.＿＿＿＿＿＿

4.＿＿＿＿＿＿

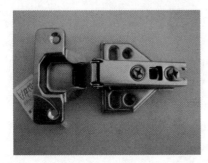

5.＿＿＿＿＿＿

6.＿＿＿＿＿＿

2.5 Find the corresponding translation for the English phrases.

1. maneuvering control
2. silicon rubber seal
3. spring force
4. girt bar
5. assist handle
6. hinge lock mechanism
7. escape slide
8. access door
9. drive mechanism
10. knot

A. 铰链锁定装置

B. 辅助手柄

C. 弹簧弹力

D. 机动控制

E. 逃生滑梯

F. 束缚杆

G. 接近门

H. 驱动装置

I. 硅胶密封垫

J. 节（航速单位）

Section 3 Aviation Translation

Translate the following English sentences into Chinese.

1. A protective cover should be put over the door frame to prevent bad weather damage to the airplane if a door keeps open for a long time.

2. Before opening operation, look to see if there is a door slide warning pennant in the door window to make sure that the escape slide is not armed.

3. Pull the exterior door control handle from the recess position to engage the door drive mechanisms.

4. An incorrectly folded strap or improperly stowed bar will interfere with the door threshold clearance.

5. If the airplane is pressurized, a properly rigged door will not unlatch.

Section 4 Aviation Writing

Situations: Yafeng Guo is an aircraft mechanic. He is inspecting the plane and finding some faults. Please help him write down fault description. Some hints of the description words, phrases & terms and the key sentences about fault location, fault description, fault solutions are offered as follows.

Key words, phrases & terms:

1. 门   door
2. 排水口   drain mast
3. 登机门   entry door
4. 货舱舱门   cargo door
5. 厨房勤务门   galley service door
6. 登机门外 / 内部手柄   entry door exterior/interior handle
7. 紧急出口   emergency exit
8. 货舱门闩   cargo door latch
9. 舱门拉动手柄   door grab handle
10. 舱门风挡遮光板   door window shade
11. 两层玻璃之间有雾气 / 水汽   fogged/moisture between panes

Key sentences:

1. 难于打开 / 关闭。   Difficult to open/close.
2. 不能电动松开。   Can not release electrically.
3. 不能完全收上。   Can not be fully retracted.
4. 前登机梯门不能在正常 / 备用方式中关闭。

   Forward airstairs door can not be closed in NORMAL/STANDBY mode.

5. 前货舱外部门锁手柄不能自动复位。

   External latch handle in the forward cargo door could not reset automatically.

Task: Yafeng Guo finds out three problems in the check. The first, cabin door unlock light was not on when the door was unlocked, the second, airstairs can not work, and the last, he could not stop motor normally after airstairs door was fully closed. Please finish the writing tasks for him.

Lesson 7 Electronic Equipment (EE) Compartment

Learning Objectives:

1. Knowledge objectives:

A. To master the major words, related terms and abbreviations about EE compartment.

B. To master the key sentences.

C. To know the LRUs of EE compartment.

2. Competence objectives:

A. To be able to understand frequently-used & complex sentence patterns, acronyms and obtain key information on aviation maintenance quickly.

B. To be able to talk about aviation or aircraft in English.

C. To be able to fill in job cards in English.

3. Quality objectives:

To be able to self-study with the help of aviation dictionaries, Internet or other resources.

Section 1 Aviation Listening and Speaking

1.1 Aviation Listening: listen to the audio and fill in the blanks with the missing words.

听力录音

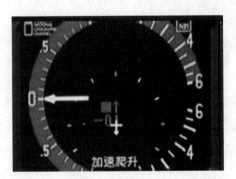

Today we will all be about the 1._____. I'm sure you've all heard the callout: "2._____, Climb" or "3._____, Descent", but where does that come from? TCAS is the 4._____ behind this voice. I'll be talking about the 5._____ Alert part first, and then go into more detail about the 6. _____ Avoidance System.

We'll go through an entire Resolution Advisory procedure, and talk about a few tricks on how to prevent a TCAS 7._____. At the end I will talk about the procedure itself, why do you turn off the 8._____ and Flight Director Bars during a TCAS event? How often do I get a TCAS Resolution Advisory and what are the minimum separation 9._____! I'll take a closer look at TCAS situations on 10._____ routes which cross arrival routes and how to reduce the risk of a

TCAS event.

1.2 Aviation Speaking: look at the pictures below and describe them in details.

Clues: well; compartment; electronic; shelf.

Section 2 Aviation Reading

课文朗读录音   译文

Pre-reading questions:

1. What does the EE compartment contain?

2. How many electronic equipment racks are there in the electronic equipment compartment?

Electronic Equipment Compartment

Introduction

Most electronic equipments are in a compartment below the main cabin floor, aft of the nose wheel well. You can enter this electronic equipment (EE) compartment through a door in the bottom of the fuselage which is aft of the nose landing gear. If necessary, the compartment can be reached in flight through a floor panel in the forward part of the passenger cabin. Modification and troubleshooting are simplified through use of easily removable shelf assemblies.

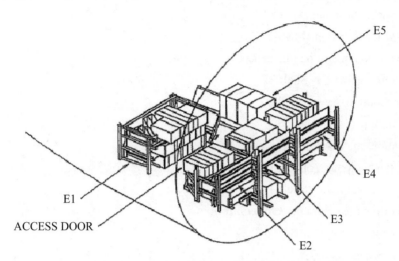

微课

Fig.7-1   Locations of Five Racks in EE Compartment

There are five electronic equipment racks in the electronic equipment compartment. They are rack E1 (in the forward part), E2 (in the left aft part) rack E3 (in the center aft part), E4 (in the right aft part) and E5 (on the right).

Equipment mounts, interconnected wiring, and accessory boxes consist of shelf assemblies. Most equipment rack shelves are cooled with air which is blown or drawn through the equipment racks. A drip shield over the racks protects the equipment from moisture condensation.

E1 Equipment Rack

The E1 has five shelves.

The E1 rack includes electronics for these functions:

➢   Autothrottle

➢   Autopilot

➢   Communication

➢   Navigation

➢   Flight control

The E1- Shelf1 includes these Line Replaceable Units (LRUs):

➢   Autothrottle computer(A/T)

➢   Flap/Slat Electronics Unit (FSEU)

➢   Flight Control Computer A (FCC A)

➢   Ground Proximity Warning Computer (GPWC)

➢   Integrated Flight Systems Accessory Unit (IFSAU)

➢   TCAS computer

Blow cools the equipment on this rack.

These LRUs are on the E1- Shelf 2:

➢ ATC transponder 1

➢ DME interrogator

➢ VOR/marker beacon receiver 1

➢ Multi-Mode Receiver (MMR)1

Blow cools the equipment on this rack.

These LRUs are on the E1- Shelf 3:

➢ Anti-skid/autobrake controller

➢ PA amplifier

➢ VHF communication transceiver 1

Blow cools the equipment on this rack.

These LRUs are on the E1- Shelf 4:

➢ Compartment overheat detector controller

➢ Flight control computer B

➢ Multi-mode receiver 2

➢ VOR/marker beacon receiver 2

Draw cools the equipment on this rack.

These LRUs are on the E1- Shelf 5:

➢ ATC transponder 2

➢ ME interrogator 2

➢ HF communication transceiver 2

Draw cools the equipment on this rack.

Fig.7-2　E1 Rack

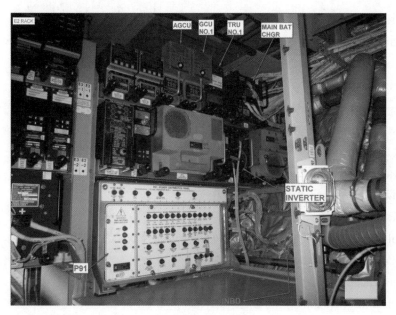

Fig.7-3   E2 Rack

E2 Equipment Rack

The E2, E3, and E4 rack contain electronics for these functions.

➢ APU

➢ Communications

➢ Common Display System (CDS)

➢ Electrical power

➢ Flight control

➢ Fire detection

➢ Window heat

The E2 has two shelves. The E2- Shelf 1 has these LRUs:

➢ APU generator control unit

➢ Battery charger

➢ Generator control unit 1

➢ Transformer rectifier

➢ Window controller 2 left front

➢ Window controller 1 right side

Blow cools the equipment on the shelf.

The Shelf E2-2 contains these LRUs:

➢ APU start power unit

➢ APU start converter unit

➢ Cabin pressure controller 1

➢ Eng and APU fire det control module

➢ Static inverter

No blow or draw cools the cooling system on this shelf.

E3 Equipment Rack

The E3-1 shelf contains these LRUs:

➢ ADF receiver 1 and 2

➢ CDS display electronics unit 1 and 2

Both blow and draw cool the equipments on the shelf.

The E3-2 shelf contains these LRUs:

➢ Audio entertainment player

➢ Digital flight data acquisition unit

➢ Engine vibration signal conditioner

➢ Engine accessory unit

➢ Low range radio altimeter 2

➢ Stall management/yaw damper 1 and 2

Draw cools the equipment on the shelf.

The E3-3 shelf contains these LRUs:

➢ ACARS management unit

➢ VHF communication transceiver 1

➢ Zone temperature controller 1 and 2

Draw cools the equipment on the shelf.

Fig.7-4　E3 Rack

Fig.7-5　E4 Rack

E4 Equipment Rack

The E4-1 shelf contains these LRUs:

➢ Air conditioning accessory units

➢ Cabin pressure controller 2

➢ SELCAL decoder

➢ Remote electronics unit.

Draw cools the equipment on the shelf.

The E4-2 shelf contains these LRUs:

- ➢ Audio multiplexer
- ➢ Auto speed brake accessory module
- ➢ Window controller 3 left side
- ➢ Window controller 4 right front
- ➢ Transformer rectifier 2 and 3
- ➢ Bus power control unit
- ➢ Generator control unit 2

Draw cools the equipment on the shelf.

E5 Equipment Rack

The equipment in the E5 rack includes electronics for there functions:

- ➢ Air Data Inertial Reference System (ADIRS)
- ➢ Flight Management Computer System (FMCS)

The E5-2 shelf has these LRUs:

- ➢ Flight management computer 1 and 2
- ➢ Air data inertial reference unit 1 and 2

Both draw and blow cools the equipment on the E5-2 shelf. The Portable Control Display Unit (CDU) connector is just below the E5-2 shelf.

Fig.7-6　E5 Rack

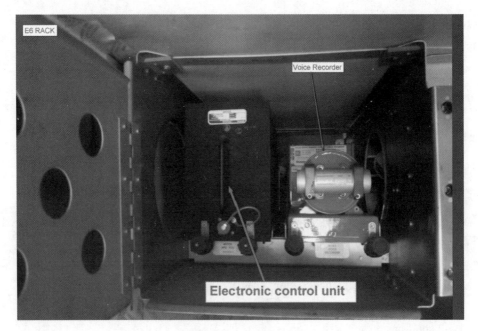

Fig.7-7　E6 Rack

An E6 equipment rack (not shown) is in the aft cargo compartment on the right side, aft of the cargo door. The rack does not have cooling air. The equipment on this rack is passively cooled.

Door Operation of the EE Compartment

Manually operate an external access door to the EE compartment from outside of the airplane.

Opening operation

Turn the handle counterclockwise to disengage the latch pins. A push-button trigger release the handle and it comes out of the door fair by spring force. Then push the door upward and slightly to the right to the first detent when the latch pins are free. Then continue to push the door to the right to the fully open detent. When the door is fully open, the roller tracks can be folded to improve access to the equipment racks. A spring catch on the end of each roller track keep the track folded or unfolded.

Closing operation

First release the spring catch and extend the roller tracks. Make sure the door frame area is clear. Then pull the door out of the fully open detent and then restrain it to slide down the tracks as gravity causes it. Release the door when it reaches the lower detent. Next turn the handle counterclockwise to retract the latch pins. If not doing this, the door will not seat in the door frame. Pull the handle to the left and the door will come out of the partially open detent. Then it slides down into the door frame. Lastly turn the handle clockwise to latch the door. Push the door handle back into its recess.

Abbreviations and Acronyms

1. Line Replaceable Units（LRUs）　　　　　　　航线可换件
2. Traffic Alert & Collision Avoidance System(TCAS)　交通警告与防撞系统
3. Digital Flight Data Acquisition Unit (DFDAU）　数字飞行数据采集组件

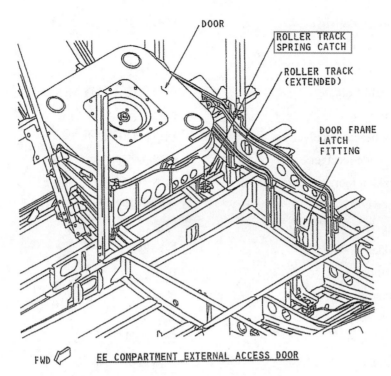

Fig.7-8　EE Compartment External Access Door

Words and Phrases

1.	electronic equipment (EE) compartment	电子设备舱
2.	mount *n.*	托架；底座
3.	wiring *n.*	线路
4.	accessory *n.*	配件；附件
5.	moisture condensation	水汽凝结物
6.	trigger *n.*	触发器
7.	latch pin	锁销
8.	detent *n.*	止动爪
9.	roller track	滚轮滑轨
10.	spring catch	弹簧锁
11.	autothrottle *n.*	自动油门
12.	autopilot *n.*	自动驾驶仪
13.	interrogator *n.*	问询机
14.	transponder *n.*	应答机
15.	multiplexer *n.*	多路调制器
16.	amplifier *n.*	（功率、声音）放大器、扩大器
17.	altimeter *n.*	高度表
18.	vibration *n.*	震动；振动
19.	decoder *n.*	解码器

20.	communication *n.*	通信
21.	navigation *n.*	导航
22.	flight control	飞行操纵
23.	replaceable *adj.*	可替代的；可置换的
24.	proximity *n.*	接近
25.	marker beacon receiver	指点信标接收机
26.	multi-mode receiver	多模接收机
27.	anti-skid/autobrake controller	防滑 / 自动刹车控制器
28.	transceiver *n.*	无线电收发器
29.	detector *n.*	探测器
30.	controller *n.*	控制器
31.	acquisition *n.*	探测
32.	altimeter *n.*	高度计
33.	inverter *n.*	反流器
34.	static inverter	静变流机
35.	converter *n.*	变流器
36.	APU Start Converter	起动变流机
37.	rectifier *n.*	整流器
38.	Transformer Rectifier Unit (TRU)	变压整流器
39.	charger *n.*	充电器
40.	bus *n.*	汇流条
41.	generator *n.*	发电机
42.	connector *n.*	连接器
43.	inertial *adj.*	惯性的
44.	portable *adj.*	手提的，便携的
45.	conditioner *n.*	调节器

Exercises

2.1 Answer the following questions according to the passage.

1. Where is the EE compartment？

2. How can we get access to the EE compartment?

3. What are the exact locations of the five racks in EE compartment?

4. What's the function of the drip shield over the racks?

5. How can we close the EE compartment door?

2.2 Translate the following terms or abbreviations into Chinese.

1. electronic equipment (EE) compartment　　2. trigger

3. mount　　　　　　　　　　　　　　　　4. latch pin

答案

5. wiring

6. detent

7. accessory

8. roller track

9. moisture condensation

10. spring catch

2.3 Give the equivalent English terms and corresponding Chinese translations according to the pictures.

1._____

2. _____

3._____

4. _____

5. _____

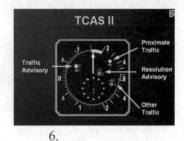

6. _____

Section 3 Aviation Translation

Translate the following English sentences into Chinese.

1. A drip shield over the racks protects the equipments from moisture condensation.

2. Equipment mounts, interconnected wiring, and accessory boxes consist of shelf assemblies. Most equipment rack shelves are cooled with air which is blown or drawn through the equipment racks.

3. Pull the handle to the left and the door will come out of the partially open detent. Then it slides down into the door frame.

4. Most electronic equipment is in a compartment below the main cabin floor, aft of the nose wheel well.

5. Next, turn the handle counterclockwise to retract the latch pins. If not doing this, the door will not seat in the door frame.

Section 4 Aviation Writing

Situations: Yafeng Guo is an aircraft mechanic. He is inspecting the plane and finding some faults. Please help him write down fault description. Some hints of the description words, phrases & terms and the key sentences about fault location, fault description, fault solutions are offered as follows.

Key words, phrases & terms:

1. 恒亮   be on steady
2. 闪亮   flash
3. 失效   fail
4. 消失   disappear
5. 显示空白   display blank
6. 不工作   do not operate/do not work
7. 不出现   do not show

Key sentences:

1. A/T 琥珀色警告信号牌灯闪亮。

   A/T amber warning annunciator light was on flashing.

2. CDU 导航状态显示页上的数据显示失效。

   Data on NAV STATUS display of CDU showed FAIL.

3. 下滑道 / 航向道偏差指针消失。

   Glide slope/Localizer deviation pointer was out of view.

4. 显示空白。Display blank.

5. 一个模式电门的电门灯不亮。

   One of the switch lights was not on for one mode switch.

6. 按压键后键不工作。

   Key did not work when pushed.

Task: Yafeng Guo finds out three problems in the check. The first, all upper switch lights were off for all mode switches, the second, EXEC key did not work when it was pushed, and the last, there was fault when the TEST switch on the navigation control panel was pushed. Please finish the writing tasks for him.

Lesson 8 Landing Gears

Learning Objectives:

1. Knowledge objectives:

A. To master the major words, related terms and abbreviations of the landing gears.

B. To master the key sentences.

C. To know the landing gear system and its components.

2. Competence objectives:

A. To be able to understand frequently-used & complex sentence patterns, acronyms and obtain key information on aviation maintenance quickly.

B. To be able to talk about aviation or aircraft in English.

C. To be able to fill in job cards in English.

3. Quality objectives:

To be able to self-study with the help of aviation dictionaries, Internet or other resources.

Section 1 Aviation Listening and Speaking

1.2 Aviation Listening: listen to the audio and fill in the blanks with the missing words.

听力录音

Today's topic will be airplane 1._____ speeds, V1, VR and V2. I'm sure many of you have heard about these speeds.

I will explain more 2._____ what is so important about V1.

By the book V1 is defined as "The speed beyond which the takeoff should no longer be aborted".

Meaning that in case you experience any trouble with your 3._____ before reaching V1, the classic example would be an 4._____ failure, you would immediately abort your take-off and would apply all necessary matters to bring the aircraft to a stop.

VR or better know as Rotate is defined as "The speed at which the 5._____ begins to apply control inputs to cause the aircraft 6._____ to pitch up, after which it will leave the ground."

The easiest way to memorize VR is, the point where the 7._____ leaves the ground vortexes are created at the 8._____ which "rotate" behind the aircraft.

V2 is defined as the takeoff 9._____ speed, the speed at which the aircraft may safely be climbed with one engine10._____.

1.3 Aviation Speaking: look at the pictures below and describe them in details.

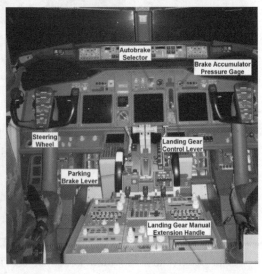

Clues: the nose & main landing gears; taxi light; actuator; retraction and extension; torsion link.

课文朗读录音　　译文

Pre-reading questions:

1. What's the function of the landing gears?

2. What do the land gears consist of?

The Landing Gears

Introduction

The 737 airplane has a tricycle type landing gear with air/oil shock struts.

The landing gear provides support for the airplane static and ground maneuvering conditions. The landing gear also reacts to airplane load forces that are generated during airplane movement. The landing gear consists of two main landing gears and a nose landing gear.

The Landing Gear System

The landing gear system consists of the following systems:

➢ The gears which support the airplane while on the ground.

➢ The gear extension and retraction system.

➢ An alternate gear extension system.

➢ Wheels and brakes for each main gear.

➢ Means for steering the airplane.

➢ Gear and door indicating and warning system.

➢ Control and operation of the gear for landing, take-off and ground movement.

The Main Landing Gear (MLG)

微课

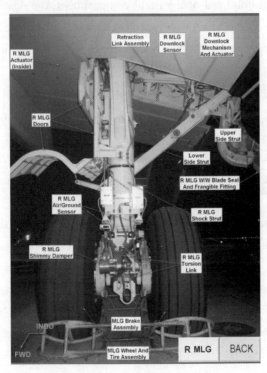

Fig. 8-1　The Right Main Landing Gear

Located inboard of each engine, aft of the rear wing spar, the main gear provides the support for the aft section of the fuselage. It absorbs impact on landing, and shocks and vibration while taxiing with a shock strut. Each main gear is hydraulically actuated to retract inboard into the fuselage. Doors and wheel well seals provide fairing with the gear retracted. Lock mechanisms and sensors assure that the main gear is down and locked or up and locked.

Two body gears and two wing gears consist of the main landing gear. Each gear is comprised of a four-wheel truck. The body gear is located at station 1463.5 in the fuselage and the wing gear is located aft of the rear wing spar inboard of the engine nacelles at station 1342.5.

Landing impact is absorbed by five air-oil shock struts, functioning primarily as air springs. Rolling vibrations and variances in runway are absorbed by the hydraulic forces within the shock strut.

Body gear trucks are steerable, providing directional control in conjunction with the nose gear in sharp turns during low speed taxiing and towing. This capability also reduces tire scrubbing in sharp turns.

The body and wing gear doors each have wheel well doors and shock strut doors. The wheel well doors are hydraulically actuated and can be closed with the gear extended or retracted. The shock strut doors are mechanically attached by linkage rods to the gear shock strut and move only when the gear is moved. All doors are of frame construction with skin paneling on the inner and outer sides. The doors close over all gear openings and fair with the fuselage contour to provide aerodynamic smoothness.

The Nose Landing Gear (NLG)

Located below the aft bulkhead of the flight compartment, the nose gear is a steerable wheel assembly and provides the support for the forward section of the fuselage and directional control on the ground. It includes a drag brace, shock strut, torsion links, a hydraulic nose gear actuator and a hydraulic lock actuator.

Landing loads are absorbed by an air-oil shock strut which consists of inner and outer cylinders. Shocks and bumps during taxi, take-off and landing are absorbed by the shock strut charged with compressed air or nitrogen.

The nose wheel steering system supplies the ground directional control by hydraulically actuated cylinders. Longitudinal stability is provided by a hinged drag brace which folds upward and aft during gear retraction.

Torsion links connected at the upper end to a steering collar and at the lower end to the shock strut inner cylinder transmit a turning moment supplied by hydraulically actuated steering cylinders. The steering is controlled by the hand wheel on the pilots' sidewall or by the rudder pedals. The hand wheel can turn the nose wheel 78 from center and the rudder pedals at full deflection can turn the nose wheel 7 from center.

The airplane can be towed forward or aft from the nose gear. The nose gear doors are clamshell type and consist of four doors that fair with the fuselage contour when closed.

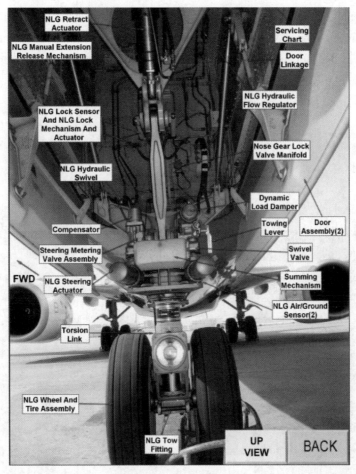

Fig. 8-2　The Nose Landing Gear

The Wheels and Brakes

The airplane is supported on 18 wheel and tubeless tire assemblies during landing, takeoff, and ground operations. Eight are on the wing gear, eight on the body gear and two are on the nose gear. Each main gear wheel (wing and body) is provided with a brake unit installed on the axle on the side nearest to the shock strut. The brakes are fitted with combination return springs and automatic adjusters. The adjusters compensate for brake wear.

When the airplane brakes are applied, the antiskid system automatically compensates for wheel skid by control of the brake pressure through the antiskid valves. A brake temperature monitoring system monitors the brake temperature and alerts the brake overheat condition of the flight crew. The tail skid system protects the lower aft fuselage if the airplane rotates too much during take-off and landing .

The Gear Extension and Retraction System

The landing gear extension and retraction systems extend and retract the landing gear hydraulically by means of a landing gear selector lever located on the center instrument panel (P2-2). When the selector lever is placed in the UP position, all three gears retract. Each main gear retracts into the fuselage and the nose gear retracts into the nose wheel well. When the selector lever is placed in the DOWN position, all three gears extend. The OFF position is the normal cruise mode and all landing gear are locked up and depressurized. A solenoid lock system restricts the selector lever to the UP position when the airplane is on the ground.

The Gear & Door Indicating and Warning System

Fig. 8-3　The Gear Indicating Panels

Landing gear control and indication components are located in the flight compartment which include:

> ➤ The landing gear selector lever and indication lights, manual extension handles.
> ➤ Parking brake lever and indicator.
> ➤ Nose gear steering control wheel.
> ➤ Antiskid control switch and inoperative indicator.
> ➤ The auto-brake control switch and disarm indicator.

Six lights above the landing gear selector lever provide indication and warning. A green light is illuminated when the respective gear is down and locked. The red light is illuminated when the landing gear is in transit or the landing gear lever and the landing gear do not agree. The red lights also serve as warning lights when the airplane is in a possible landing configuration and the gear is not down and locked.

Three red manual extension handles are located on the floor of the flight compartment. Just aft of the first officer's station, for manual extension of the landing gear in case of hydraulic malfunction in hydraulic system "A". The manual extension handles are independently operated for each gear. The landing gear selector lever should be in the OFF position for manual extension.

Components

The landing gear system has these main components:

➢ Control lever assembly

➢ Manual extension mechanism

➢ Transfer valve

➢ Selector valve

➢ Main landing gear (2)

➢ Nose landing gear

➢ Shimmy damper

➢ Proximity Switch Electronics Unit (PSEU)

➢ Landing gear panel

➢ Auxiliary landing gear position lights

Fig. 8-4   PSEU

The pressure to the landing gear extension and retraction is normally supplied by hydraulic system A. Hydraulic system B only supplies pressure for retraction. The landing gear transfer valve receives electrical signals from the Proximity Switch Electronics Unit (PSEU). It changes the pressure source of the landing gear from hydraulic system A to hydraulic system B. Move the landing gear control lever assembly to control landing gear extension and retraction. This lever moves the selector valve through cables. The selector valve also gets an electrical input from the manual extension system. This operates a bypass valve in the selector valve to connect the landing gear retraction to the hydraulic system return. This lets the manual extension system extend the landing gear.

Landing gear lights show the position of the landing gear. The PSEU receives landing position signals from sensors on the landing gear. The normal and auxiliary lights are controlled by the PSEU.

The pressure for nose wheel steering comes from the nose landing gear extension pressure only. Hydraulic system A normally supplies pressure to the nose gear steering through the landing gear control system.

The Landing Gear Control Operation

The landing gear control system also provides normal or alternate hydraulic pressure to these systems:

➢ Main landing gear shimmy damper.

➢ Gear retract brake system.

Install a downlock pin into the main and the nose landing gear to make sure an outside force does not unlock the landing gear. There is one main landing gear downlock pin for each main landing gear and one nose landing gear downlock pin for the nose landing gear. The downlock pin installs in the MLG downlock strut and the nose landing gear downlock pin installs in the NLG downlock pin hole. Install the ground locks in all landing gear very carefully for an accidental retraction can cause injury to persons and damage to equipment.

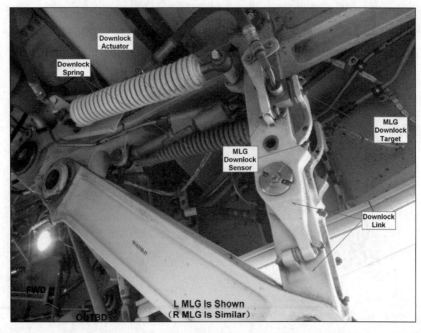

Fig. 8-5　Down lock Pin of the Main Landing Gear (L MLG)

Abbreviations and Acronyms

1. Main Landing Gear (MLG)   主起落架

2. Nose Landing Gear (NLG)   前起落架

3. Proximity Switch Electronics Unit (PSEU)   接近电门电子组件

Words and Phrases

1. main gear   主起落架

2. nose gear   前起落架

3. mount *v.*   安装

4. cruise mode   巡航模式

5. solenoid lock   电磁锁

6. taxi *v.*   滑行

7. drag brace   阻力撑杆

8. torsion link   扭力臂

9. trunnion *n.*   耳轴；枢轴

10. tow lug   牵引环

11. steering collar   转向套管环

12. turning moment   转矩；转弯力矩

13. tail skid   尾撬，尾部滑撬

14. transfer valve   转换活门

15. shimmy damper   减摆器

16. bypass valve   旁通活门

17. downlock pin   下位锁销

Exercises

答案

2.1 Answer the following questions according to the passage.

1. What is the function of the gear?

2. What does the land gear consist of?

3. What provides the support for the aft section of the fuselage?

4. Where are the landing gear control and indication components?

5. Which system supplies pressure to the landing gear extension and retraction?

2.2 Translate the following terms or abbreviations into Chinese.

1. control lever assembly   2. bypass valve

3. the main landing gear   4. nose landing gear

5. cruise mode   6. downlock pin

7. tow bar   8. shock strut

9. torsion link   10. transfer valve

2.3 Give the equivalent English terms and corresponding Chinese translations according to the pictures.

1._____

2. _____

3._____

4. _____

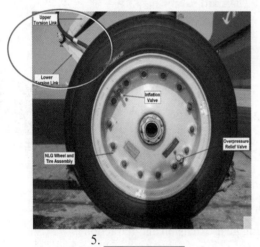

5. _____

6. _____

Section 3 Aviation Translation

Translate the following English sentences into Chinese.

1. Torsion links connected at the upper end to a steering collar and at the lower end to the shock strut inner cylinder transmit a turning moment supplied by hydraulically actuated steering cylinders.

2. The doors close over all gear openings and fair with the fuselage contour to provide aerodynamic smoothness.

3. Body gear trucks are steerable, providing directional control in conjunction with the nose gear in sharp turns during low speed taxiing and towing.

4. Each main gear wheel (wing and body) is provided with a brake unit installed on the axle on the side nearest the shock strut.

5. Install the ground locks in all landing gear very carefully for an accidental retraction can cause injury to persons and damage to equipment.

Section 4 Aviation Writing

Situations: Yafeng Guo is an aircraft mechanic. He is inspecting the plane and finding some faults. Please help him write down fault description. Some hints of the description words, phrases & terms and the key sentences about fault location, fault description, fault solutions are offered as follows.

Key words, phrases & terms:

1.	机轮	wheel
2.	刹车	brake
3.	起落架	landing gear
4.	轮舱	wheel well
5.	主轮	main landing gear wheel
6.	外起落架轮子	outboard main wheel
7.	抱、拖 / 锁死	grab, drag/lock
8.	起落架轮胎	serve landing gear tire
9.	放出 / 收上	extended/retracted
10.	前起落架支柱	nose landing gear shock strut
11.	润滑前起落架	lubricate nose landing gear

Key sentences:

1. 在轮舱内造成很大的噪音。Made loud noise in the wheel well.

2. 更换左外起落架轮子。Replace L/H outboard main wheel.

3. 右前轮无法正常充气 The R/H of the forward tire can not be charged normally.

4. 停止过程中过热。Overheated during stop.

5. 向左 / 右偏。Pulled to the left/right.

6. 起落架手柄在中立（OFF）位时，所有起落架伸出、前起落架、右起落架及左起落架绿灯亮。

   All gear extended with landing gear lever at OFF, NOSE GEAR, RIGHT GEAR, and LEFT GEAR green lights were on.

7. 起落架手柄在放下（DN）/ 收上（UP）位时，左起落架、右起落架及前起落架红灯和绿灯亮。

LEFT GEAR, RIGHT GEAR, and NOSE GEAR red lights and green lights were on with landing gear lever at DN/UP.

8. 起落架手柄在飞行中，不能移到收上位；当使用锁定超控时，起落架手柄可自由移至收上位。

Landing gear lever can't be moved to the UP position in flight, it's free to be moved to the UP position when the lock override was used.

9. 当飞机在地面时，前起落架红灯亮。

NOSE GEAR red lights were on when airplane was on ground .

10. 主起减震支柱的勤务。Serve main landing gear shock strut.

11. 调节并测试主起落架安全电门。Adjust/test main landing gear safety switch.

起落架位置灯显示不正确。Landing gear position light displayed incorrectly.

12. 润滑主起落架。Lubricate main landing gear assembly.

13. 更换左外侧主轮。Replace L/H outboard main landing gear wheel.

Task: Yafeng Guo finds out three problems in the check. The first, landing gear position light can not be displayed correctly, the second, to lubricate main landing gear assembly, the last, to replace left outboard main landing gear wheel.

Lesson 9 The Powerplant

Learning Objectives:

1. Knowledge objectives:

A. To master the major words, related terms and abbreviations about powerplant.

B. To master the key sentences about powerplant.

C. To know the basic structure of the CFM56-7B.

2. Competence objectives:

A. To be able to understand frequently-used & complex sentence patterns, acronyms and obtain key information on aviation maintenance quickly.

B. To be able to talk about aviation or aircraft in English.

C. To be able to fill in job cards in English.

3. Quality objectives:

To be able to self-study with the help of aviation dictionaries, Internet or other resources.

Section 1 Aviation Listening and Speaking

1.1 Aviation Listening: listen to the audio and fill in the blanks with the missing words.

听力录音

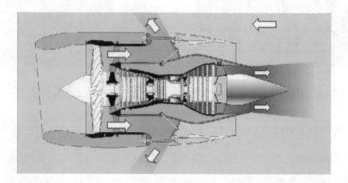

1._____ is used to slow down the aircraft on the runway after touchdown. There are three main 2._____ which slow down the airplane on the runway: Primary 3._____ with disc or carbon brakes similar to your car. Secondary with reverse thrust. Third aerodynamically braking with the 4._____.

So we have two words in reverse thrust. "Reverse" cause the 5._____ output is being guided into the reverse direction and as you might not know we apply "thrust" in order to increase the braking action.

1.2 Aviation Speaking: look at the pictures below and describe them in details.

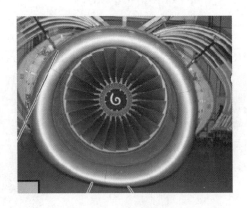

Clues: thrust; heart; engine; EEC; fan; CFM.

Section 2 Aviation Reading

课文朗读录音   译文

Pre-reading questions:

1. What is the function of the airplane powerplant?
2. What is the engine type used in B737-800?

The Powerplant

The powerplant provides the thrust for flight and also supplies thrust reversers power in stopping the airplane. It also provides bleed air for air conditioning, pressurization, anti-icing, and drives a gearbox used for electrical power and hydraulic power. The 737 airplane are powered by two wing-mounted CFM 56-7 B engines, and the engine buildup is identical for left or right engine. The engine is the heart of an airplane, which makes us know how important the engine is. Different type engine may be used in different BOEING airplane. Next let's take CFM56-7B, the engine in

B737-800, as an example to get a general idea of a turbofan engine.

Fig.9-1　Turbofan Engine

The CFM56-7B is a high bypass ratio, a two-spool, axial flow turbofan engine. The engine fan diameter is 61 inches (1.55 meters). The bare engine weight is 5257 pounds (2385 kilograms). It is supported by the wing pylon and streamlined by cowlings.

微课

Fig.9-2　CFM56-7B Engine

The engine cowling is a cover that contains and gives protection to the engine components. The cowling also controls the airflow through and around the engine. The inlet cowl supplies a smooth airflow into the engine. The fan cowl is aft of the inlet cowl. The fan cowl attaches to the fan cowl support beam on the engine. The thrust reverser is aft the fan cowl. The thrust reverser attaches to the thrust and opens for maintenance. The turbine exhaust system is aft of the thrust reverser. The system has three parts: primary nozzle, forward center-body and aft center-body.

Engine Components

① inlet   ② fans   ③ core module   ④ LPT module

Fig.9-3   Engine Components

The engine is made up of four main sections: compressor section, combustion section, turbine section, and accessory drives.

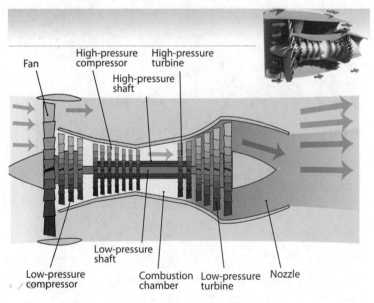

Fig.9-4   Engine Sections

The engine components include fan, Low Pressure Compressor (LPC), High Pressure Compressor

(HPC), Combustion chamber, High Pressure Turbine (HPT), Low Pressure Turbine(LPT), exhaust nozzle and accessory drive.

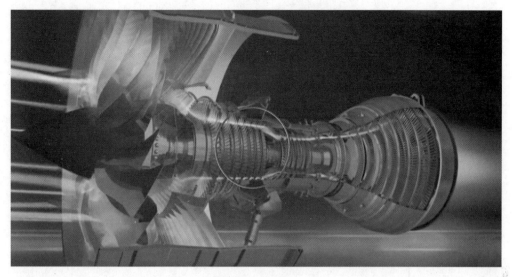

Fig.9-5   Compressor

Fig.9-6   Exhaust Nozzle

Fan and booster

The fan and booster is a four-stage compressor. The fan, located on the front of the engine, is the first component of a turbofan engine. The main function of the fan is to increase the speed of the air. The large spinning fan sucks in large quantities of air. Air is sucked into the intake by the fan blades. Most blades of the fan are made up of titanium. A splitter fairing divides the air into primary airflow and secondary airflow.

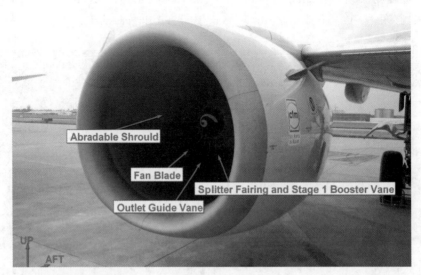

Fig.9-7　Fan

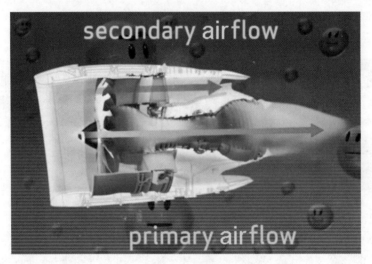

Fig.9-8　Booster

The primary airflow passes through the inner portion of the fan blades and is directed into a booster. The airflow then enters a High Pressure Compressor (HPC). After the primary airstream has been compressed by the LPC and HPC, it goes to a combustor. Mixed with fuel and ignited, the gas flow in the annular combustion chamber increases the HPC discharge air velocity to drive the high pressure turbine (HPT) and low pressure turbines (LPT), and exits through the core exhaust nozzle.

The secondary airflow passes through the outer portion of the fan blades, the Outlet Guide Vanes (OGV's) and exits through the fan discharge duct, producing approximately 80% of the total thrust. It also plays a role in the thrust reverser system.

The bypass ratio of the engine is about 5:1. This means that approximately five times more air goes through the fan duct than through engine core.

Low Pressure Compressor (LPC)

The LPC is mounted on the rear of the fan disk. It is a three-stage compressor. The three-stage of blades are numbered 2 to 4, as the first stage of LPC is the fan blade. The LPC stator and rotor assembly provides the initial compression of the ambient (inlet) air which is then directed to the HPC. The LPC rotor is driven by the LPT.

High Pressure Compressor (HPC)

The HPC is aft of the LPC. It is a nine-stage compressor. The purpose of the HPC is to further compress air from the LPC and sends it to the combustor. The HPC also supplies bleed air for the aircraft pneumatic system and the engine air system. The HPC is on the same shaft as HPT. The HPT turns the N2 shaft.

Combustion chamber

The front face of the combustor is attached to the rear of the HPC. The combustion chamber is a short annular structure. In the combustor, the high pressure air from the HPC is mixed with the fuel from the fuel nozzles. There are as many as 20 nozzles to spray fuel to the airflow. This mixture of air and fuel is ignited by 2 igniter plugs to make hot expanding gases, in order to produce the necessary energy to drive the turbine rotors. Residual energy is converted into thrust.

High Pressure Turbine (HPT)

Aft of the combustion section is the HPT. The HPT is a single-stage turbine. The hot gases coming out of the combustor goes into the HPT. The HPT changes the energy of the hot gases into mechanical energy. The HPT uses this mechanical energy to turn the HPC rotor and the accessory drive.

Low Pressure Turbine (LPT)

The LPT is aft of the HPT. It is a four-stage turbine. It changes the energy of the hot gases into echanical energy. The LPT uses this mechanical energy to turn the fan and booster rotor.

Exhaust nozzle

The exhaust nozzle is the exhaust duct of the engine. This is the engine part which actually produces the thrust for the airplane. The energy depleted airflow that passed the turbine, in addition to the colder air that bypassed the engine core, produces a force when exiting the nozzle that acts to propel the engine, and therefore the engine, forward.

Accessory drive

The Accessory GearBox (AGB) is on the left side of the engine, on the fan inlet case. The accessory drive has three gear boxes and two shafts. The three gear boxes are Inlet Gear Box (IGB), Transfer Gear Box (TGB) and Accessory Gear Box (AGB). The two shafts are Radial Drive Shaft (RDS) and Horizontal Drive Shaft (HDS). The inlet and transfer gearboxes transfer energy from the N2 shaft to the accessory gearbox and the accessory gearbox holds and turns the engine accessories. The N2 shaft turns the radial drive shaft through the inlet gearbox. The radial drive shaft turns the horizontal drive shaft through the transfer gearbox. The horizontal drive shaft turns the accessory gearbox.

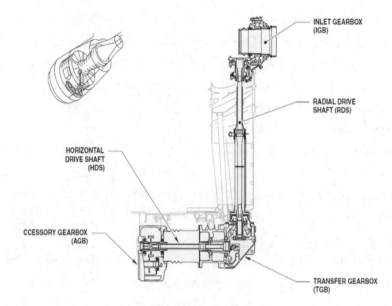

Fig.9-9　Accessory Drive Design

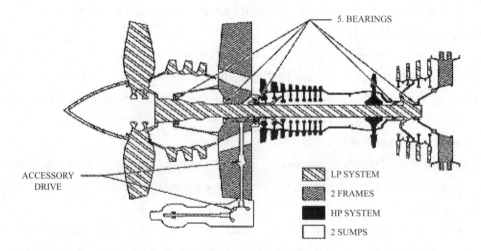

Fig.9-10　Accessory Drive

You get access to the AGB and the accessories when you open the left fan cowl. The engine and airplane accessories located on the front of the accessory gear box are as follows: EEC alternator, N2 sensor, hand cranking pad, engine air starter, Integrated Drive Generator (IDG), hydraulic pump. The engine accessories located on the aft of the accessory gear box are as follows: fuel pump package (fuel pumps, HMU, and main oil/fuel heat exchanger), lubrication unit, scavenge oil filter.

Use the hand cranking pad to turn the N2 rotor during borescope inspection.

Main engine bearings

There are five main engine bearings holding the N1 shaft and the N2 shaft. Ball bearings absorb the axial and the radial loads from the shafts. Roller bearings absorb only radial loads. The main engine bearings are in two sump cavities named the forward sump and the rear sump.

No. 1 & 2 bearings hold the front of the N1 shaft. One ball bearing and one roller bearing are the number 3 bearing assembly. Both No. 3 bearings hold the front of the No. 2 shaft. No. 4 bearing holds the rear of the No. 2 shaft. No. 5 bearing holds the rear of the N1 shaft. No. 1, 2, and 3 bearings are in the forward sump. No. 4 & 5 bearings are in the rear sump.

The fan and Low Pressure Compressor (LPC) are driven, via the low pressure shaft (N1), by the Low Pressure Turbine (LPT). The High Pressure Compressor (HPC) is driven, via the high pressure shaft (N2), by the High Pressure Turbine (HPT).

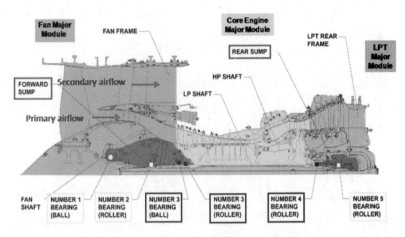

Fig.9-11　Main Engine Bearings

Engine Control

The engine control includes the engine start control, the forward thrust control and the reverse thrust control.

The start lever control fuel on & off and ignition during engine start. It switches on the ignition and the fuel supply to the combustion chamber. The master switches on the engine start panel control the engine start sequence. The fuel is supplied via a high pressure fuel shut-off valve and a low pressure fuel shut-off valve. The low pressure fuel shut-off valve is usually an electric motor driven valve. The position of it is indicated by the fuel valve closed light. The high pressure fuel shut off valve is mechanically controlled. The HP valve opens when the engine start lever is moved up and it closes when the lever is moved down. The position of the HP fuel shut-off valve is indicated by the position of the engine start lever.

The forward thrust lever controls the thrust of an engine between idle and take-off thrust. The lever informs the fuel control unit how much thrust is requested by the pilot.

When the reverse thrust lever is activated, the deflector doors are deployed and the airflow is deflected to give reverse thrust. When the reverse thrust lever is pushed down, the deflector doors move to the stowed position and the engine delivers forward thrust. The control levers help to set the required thrust and the reverse latching levers are used with the control levers to give reverse thrust.

Monitoring and Maintaining

The concept of "On Condition Maintenance" is used in the CFM56-7B engine maintenance. The engine has no periodic overhaul schedules and can remain installed under the wing until something important occurs, or when lifetime limits of parts are reached.

Engine Hazard

The engine hazard means the potential danger of working around the engine. The hazard around the engine in operation include inlet suction, exhaust heat, exhaust velocity, and noise. Keep away from the inlet and exhaust hazard area when the engine is running. Engine entry corridors are between the engine inlet hazard areas and the exhaust hazard areas. If the wind is over 25 knots, increase the inlet hazard area by 20%.

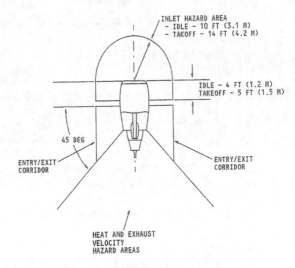

Fig.9-12　Power Plant Engine Hazard

Inlet suction

Suction at the inlet of an engine can pull people and objects into the engine. Use the engine entry corridor to go near an engine in operation. Make sure the engine is at idle power and you can speak with people in the flight compartment when you have to go near an engine in operation. Loose objects, such as rags, eyeglasses, hats, etc., can cause engine damage if they are sucked into the inlet cowl. Any loose object around the engine must be attached or removed before you work around the engine.

At idle power, the inlet hazard area is a 10 feet（3.1m）radius around the inlet.

Exhaust heat

The exhaust temperature and exhaust velocity can remain very high for hundreds of feet behind the engine. The heat and velocity of the engine exhaust can cause injury to personnel and damage to equipment.

Exhaust velocity

Exhaust velocity is very high for a long distance behind the engine, which can cause damage to personnel and equipment.

Engine noise

The engine noise can cause temporary and permanent loss of hearing. You must wear cup-type ear protector when going near an engine in operation.

Abbreviations and Acronyms

1. Low Pressure Compressor (LPC)   低压压气机
2. High Pressure Compressor (HPC)   高压压气机
3. Low Pressure Turbine (LPT)   低压涡轮
4. High Pressure Turbine (HPT)   高压涡轮
5. Outlet Guide Vane (OGV)   出口导向叶片
6. Accessory Gear Box (AGB)   附件齿轮箱
7. Inlet Gear Box (IGB)   进口齿轮箱
8. Transfer Gear Box (TGB)   传输齿轮箱
9. Radial Drive Shaft (RDS)   径向传动轴
10. Horizontal Drive Shaft (HDS)   水平传动轴

Words and Phrases

1. fan *n.*   风扇
2. compressor *n.*   压气机
3. combustor *n.*   燃烧室
4. thrust *n.*   推力
5. reverse thrust *n.*   反推
6. hydraulic *adj.*   液压的
7. pneumatic *adj.*   气动的
8. pressurization *n.*   增压
9. identical *adj.*   相同的
10. bypass ratio   涵道比
11. turbine *n.*   涡轮
12. turbofan *n.*   涡扇
13. cowling *n.*   整流罩
14. booster *n.*   压气机
15. shaft *n.*   轴
16. accessory drive   附件传动
17. borescope inspection   孔探检查
18. nozzle *n.*   喷管；喷嘴
19. fuel *n.*   燃油
20. ignite v.   点燃
21. bleed air   引气
22. approximately adv.   大约地
23. bearing *n.*   轴承
24. overhaul *n.*   大修
25. hazard *n.*   危险

26. in operation	运转中
27. suction *n.*	吸力
28. velocity *n.*	速度
29. knot *n.*	节

答案

Exercises

2.1 Answer the following questions according to the passage.

1. What are the functions of powerplant?

2. What are the main features of CFM56-7B?

3. How many components do the engine have? What are they?

4. Does the engine have periodic overhaul schedules?

5. What shall we do if we have to go near an engine in operation?

2.2 Translate the following terms or abbreviations into Chinese.

1. high pressure compressor (HPC)   2. thrust

3. low pressure turbine (LPT)   4. bearing

5. accessory gearbox (AGB)   6. overhaul

7. horizontal drive shaft (HDS)   8. bypass ratio

9. accessory drive   10. cowling

2.3 Give the equivalent English terms and corresponding Chinese translations according to the pictures.

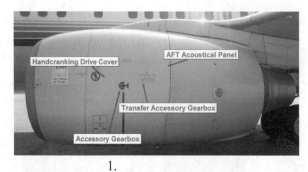

1._____

2. _____

3._____

4. _____

Fuel Nozzle Filter

5. _____

6. _____

Section 3 Aviation Translation

Translate the following English sentences into Chinese.

1. The primary airflow passes through the inner portion of the fan blades and is directed into a booster (LPC).

2. The purpose of the HPC is to further compress air from the LPC and send it to the combustor.

3. In the combustor, the high pressure air from the HPC is mixed with the fuel from the fuel nozzles.

4. The LPT uses this mechanical energy to turn the fan and booster rotor.

5. The hazard around the engine in operation includes inlet suction, exhaust heat, exhaust velocity, and noise.

Section 4 Aviation Writing

Situations: Yafeng Guo is an aircraft mechanic. He is inspecting the plane and finding some faults. Please help him write down fault description. Some hints of the description words, phrases & terms and the key sentences about fault location, fault description, fault solutions are offered as follows.

Key words, phrases & terms:

1. 风扇   fan
2. 气缸   cylinder
3. 发动机   engine
4. 尾喷管   nozzle
5. 进气罩   inlet cowl
6. 释压门   pressure relief door
7. 滑油箱   oil tank
8. 风扇罩   fan cowl
9. 动力装置   powerplant

10.	风扇叶片	fan blade
11.	发动机架	engine mount
12.	附件齿轮箱	accessory gearbox
13.	传动齿轮箱	transfer gearbox
14.	转子，叶轮	spinner
15.	发动机短舱	nacelle
16.	发动机停车	engine shutdown
17.	发动机熄火	engine flame out
18.	发动机慢车位	idle
19.	发动机驱动泵	EDP(engine driven pump)
20.	电动马达驱动泵	EMDP(electric motor-driven pump)
21.	进气整流罩	nose cowl
22.	发动机安装架	engine mount

Key sentences:

1. 左发进气道防磨层破损。Rubber of the inlet cowl of the L/H engine has damaged.
2. 左发遭鸟击，风扇叶片的平垫块有 5 个受损。Five plain blocks of the fan vane of the L/H engine were damaged because of bird impacting.
3. 拆除并安装风扇叶片。Remove and install the fan blades.
4. 可变静子叶片反馈钢索的调节。Adjust VSV feedback cable.
5. 更换左侧可变静子叶片作动筒。Replace the L/H side VSV actuator.
6. 发动机的拆装。Remove and install engine.
7. 涡轮风扇活门的拆装。Remove and install turbofan valve.

Task: Yafeng Guo finds out three problems in the check. The first, rubber of the inlet cowl of the right engine has damaged, the second, the plain blocks of the fan vane of the right engine were damaged because of bird impacting, the last, to remove and install the front and rear spinner cones.

Lesson 10 Auxiliary Power Unit (APU)

Learning Objectives:

1. Knowledge objectives:

A. To master the major words, related terms and abbreviations about APU.

B. To master the key sentences about APU.

C. To know the functions of APU.

2. Competence objectives:

A. To be able to understand frequently-used & complex sentence patterns, acronyms and obtain key information on aviation maintenance quickly.

B. To be able to talk about aviation or aircraft in English.

C. To be able to fill in job cards in English.

3. Quality objectives:

To be able to self-study with the help of aviation dictionaries, Internet or other resources.

Section 1 Aviation Listening and Speaking

1.1 Aviation Listening: Listen to the audio and fill in the blanks with the missing words.

听力录音

What is ram air turbine, or better known as the RAT. We'll be looking at some basic system 1._____ on the Airbus Hydraulics (the green, blue and yellow system) and then dive straight in to an ELEC EMER CONFIG. You'll see how the ram air turbine is deployed and I'll 2._____ more 3._____ about the ram air turbine duties. We'll use a AC BUS 1&2 FAULT as a primary example and 4._____ the RAT 5._____ electricity. And the second scenario will be a PUAL Engine failure, and how the RAT 6._____ electrical power plus hydraulic 7._____ to be able to control the aircraft in such an emergency event. This is a 8._____ of the Airbus A320 Ram Air Turbine. Future videos 9._____ we'll 10._____ the hydraulic system in more detail are in the making.

1.2 Aviation Speaking: look at the pictures below and describe them in details.

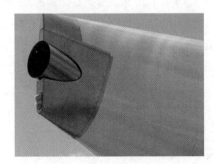

APU 辅助动力装置

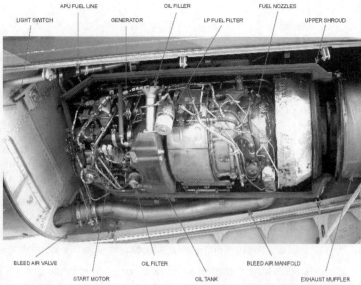

LIGHT SWITCH APU FUEL LINE GENERATOR OIL FILLER LP FUEL FILTER FUEL NOZZLES UPPER SHROUD

BLEED AIR VALVE START MOTOR OIL FILTER OIL TANK BLEED AIR MANIFOLD EXHAUST MUFFLER

Clues: Auxiliary power; turbine engine; in the tail cone.

Section 2 Aviation Reading

课文朗读录音   译文

Pre-reading questions:

1. Where is the APU located?
2. What's the function of APU?

Auxiliary Power Unit

Introduction

An Auxiliary Power Unit (APU) is a small,multipurpose gas turbine engine. It is a device that provides energy for functions other than propulsion. It is located in the tail cone of the airplane. The APU applies electrical power and pneumatic power in the ground and in flight. With it on board, electricity could be generated without keeping a main engine running,air-conditioning is available without an external air supply,and engines could be started at the touch of a button—independent of a ground cart.

The APU has three sections: the power section, the bleed section, and the accessory gearbox section. The power section of the APU drives the compressor and the gearbox. This gearbox drives all APU accessories like the fuel pump, the oil pumps, the cooling fan, and the AC-generator. On very large APU, two AC-generators are on the gearbox.

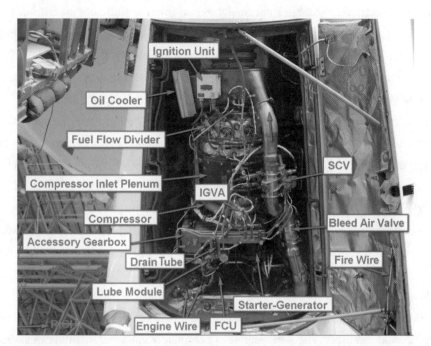

微课

Fig.10-1   Detailed Components of an APU

Location

The APU is located in the tail cone and is isolated from the fuselage by a firewall. This location is several feet higher than the fuel tanks,thereby necessitating a pressurized fuel supply.To satisfy this requirement during starts,the AC boost pump is supplemented by a parallel DC pump.If AC pumps are off,the DC pump is energized automatically when the APU starts one second initiated and continues to operate until the AC pump is activated.

Fig.10-2   Location of an APU

Primary Function

The APU has the following functions. Firstly, it can provide compressed air for engine starting, then it can provide compressed air to the air conditioning system on the ground or in flight, and thirdly it can provide electrical power to the airplane electrical system on the ground or in–flight.

The primary purpose of the APU is to provide power to start the main engines. Turbine engines must be accelerated to a high rotational speed to provide sufficient air compression for self-sustaining operation. Smaller jet engines are usually started by an electric motor, while larger engines are usually started by an air turbine motor. Before the engines are to be turned, the APU is started, generally by a battery or hydraulic accumulator. Once the APU is running, it provides power (electric, pneumatic, or hydraulic, depending on the design) to start the aircraft's main engines.

The APU generator supplies up to the electrical power to the airplane electrical system. Some APU generally produces 115 V alternating current (AC) at 400 Hz (rather than 50/60 Hz in mains supply), to run the electrical systems, others can produce 28 V direct current (DC). APU can provide power through single phase or three-phase systems.

The APU load compressor supplies pressed air for the pneumatic systems. APU drives a 90-kilowatt generator and a centrifugal compressor. The compressor supplies air to start the main engines and support the environmental control system, warming or cooling the cabin for passenger comfort.

APU Engine

The APU engine is located in the aft end of the fuselage, behind the pressure bulkhead and below the horizontal stabilizer. The APU engine is completely enclosed in a titanium shroud. The air inlet door is located on the right side of the aft fuselage for air entrance. The airflow is split into two paths, one for the engine and the other for the accessories cooling.

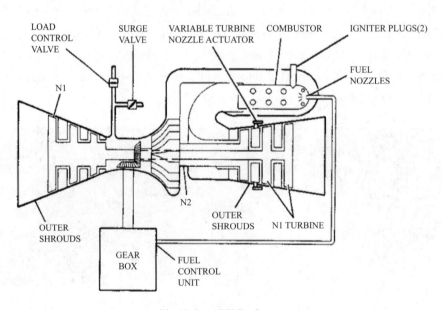

Fig.10-3　APU Engine

APU Fuel System

The APU fuel system consists of a fuel valve, control unit, solenoid valve, filter, heater and control valve. The fuel valve allows fuel supply from the tank to the APU engine. This valve operation is controlled by the APU switch on P5 panel. It is opened by placing the APU switch to the ON position and closed by the switch to the OFF position. The valve can be operated manually without power by the override handle.

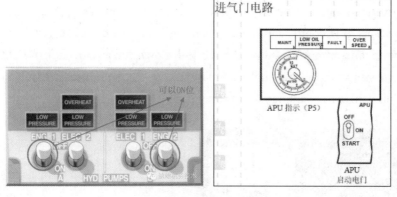

Fig.10-4　APU Fuel System

The fuel from the tank is delivered to the fuel control unit through the fuel valve. The fuel control unit supplies and regulates the fuel to the combustion chamber. The regulated fuel flow controls the acceleration of the APU engine during the starting operation.

APU Ignition & System

The APU ignition and starting system consists of a start switch, starter, start relay, ignition exciter, igniter plug, and control circuit. The starter provides initial power for rotating the APU engine compressor, turbine, and engine–driven accessories to a speed high enough to obtain good airflow for combustion. The ignition exciter provides the high voltage required to produce the spark at the igniter plug.

APU Start System

The start system interfaces with the sequencing oil pressure switch, speed switch, APU control unit and APU switch. During APU start, the starter is energized and APU rotates. The ignition exciter is energized by the action of the sequencing oil pressure switch (this switch ensures that combustion cannot be initiated during an APU start until a minimum lubricating oil pressure has been built up during APU start, with increasing oil pressure, this switch closes at 4 (Pound per square inch, gauge) psig oil pressure and provides 28 volt DC power to the fuel solenoid valve and the ignition exciter). The starter and ignition circuits are de-energized by the action of the speed switch during APU engine acceleration. During normal engine operation, the exciter remains de-energized.

The Indications & Controls

The indications and controls for the APU are in three areas. They are on the flight deck, attached to the nose strut, and in the aft equipment center.

The APU control switch, APU run light, and APU fault light are located on the P5 overhead panel. Additional information is showed in the Engine Indication and Crew Alerting System (EICAS). If there is an APU protective shut down, an EICAS advisory message "APU FAULT" appears, and the amber fault light comes on. The APU fire switch is on the center control stand.

Attached to the nose strut is the APU remote control panel (P62 panel). The APU five warning light, the APU fire warning horn, the APU fire bottle discharge switch, and the APU remote shut down switch are on this panel. This switch is for emergency shutdown only.

The APU battery, battery charger, and the APU control unit are in the aft equipment center on the E6 rack. The APU control unit controls and monitors all critical parameters in operations. The control unit starts an automatic protective shut down for any of 16 different reasons. The APU control unit has Built-In Test Equipment (BITE). Protective shut-down faults and failed Line Replaceable Unit (LRU) are identified on the front panel.

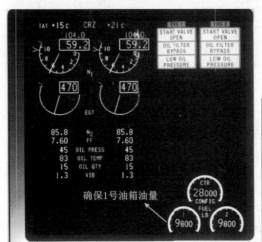

Fig.10-5   APU Indications & Controls

APU Driven Generator

The APU driven generator is mounted on the forward face of the APU accessory drive gearbox. The APU generator produces the AC power required by the airplane systems primarily for ground operation. The generator can also be used as a backup generator in flight. APU is used to supply 60 KVA of electrical power on the ground and used as AC power source of 50 KVA in flight.

The two APU generator breakers connect 115 volt AC, 400 hertz, 3-phase power to the distribution system. The APU generator breakers are located in the bottom of the P6 panel in the flight compartment. An internal spring assists opening and holds it in the open position. The breaker has six main 115 volt AC contacts and 20 auxiliary contacts. The auxiliary contacts are used to control the positions of APU generator breakers and external power contactors, and for indicating lights in the flights in the flight compartment. The closing and tripping signals for the breakers are from the P5-4 panel. The breaker contacts position is indicated by the blue APU GEN OFF BUS light on the P5-4 panel.

Abbreviations and Acronyms

1. Auxiliary Power Unit (APU)   辅助动力装置
2. Alternating Current (AC)   交流电

3.	Direct Current (DC)	直流电
4.	Engine Indication &Crew Alerting System (EICAS)	发动机指示和机组警告系统
5.	Built-In Test Equipment (BITE)	内置测试设备
6.	Line Replaceable Unit (LRU)	航线可换件
7.	Kilo Volt-Ampere (KVA)	千伏安

Words and Phrases

1.	turbine *n.*	涡轮
2.	propulsion *n.*	推进
3.	tail cone	尾部整流器
4.	pneumatic *adj.*	气动的
5.	bleed section	输出部分
6.	accessory gearbox	附件齿轮箱
7.	boost pump	增压泵
8.	self-sustaining *adj.*	自给的；自立
9.	hydraulic accumulator	蓄压器
10.	single phase	单相
11.	load compressor	负载压缩机
12.	centrifugal compressor	离心式压缩机
13.	titanium shroud	钛防护罩
14.	valve *n.*	活门
15.	solenoid valve	电磁阀
16.	filter *n.*	过滤器
17.	override handle	操纵柄
18.	regulate *v.*	调节
19.	combustion chamber	燃烧室
20.	start switch	起动开关
21.	starter *n.*	起动器
22.	start relay	起动继电器
23.	ignition exciter	点火励磁器
24.	igniter plug	点火器插头
25.	control circuit	控制电路
26.	compressor *n.*	压气机
27.	hertz *n.*	赫兹
28.	lubricating oil pressure	油润滑压力

Exercises

2.1 Answer the following questions according to the passage.

1. How many sections do APU consist of?

答案

2. What's the function of every section of APU?

3. How many components does the APU fuel system include?

4. What does the APU ignition and system consist of ?

5. Where are the indications and controls for the APU?

6. Where is the APU driven generator located?

7. What's the function of the APU driven generator?

2.2 Translate the following terms or abbreviations into Chinese.

1. firewall

2. boost pump

3. the air conditioning system

4. hydraulic accumulator

5. fuel valve

6. solenoid valve

7. start switch

8. ignition exciter

9. igniter plug

10. lubricating oil pressure

2.3 Give the equivalent English terms and corresponding Chinese translations according to the pictures.

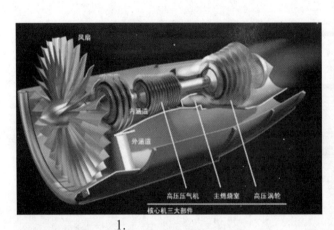

1._____

2._____

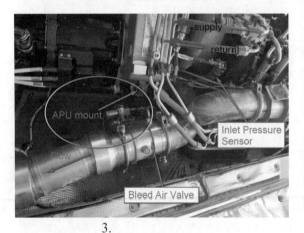

3._____

4._____

5. _____   6. _____

2.4 Find the corresponding translation for the English phrases.

1. auxiliary power unit (APU)   A. 空调系统
2. the air conditioning system   B. 滑油润滑压力
3. lubricating oil pressure   C. 辅助动力装置
4. solenoid valve   D. 点火励磁器
5. fuel valve   E. 液压蓄能器
6. firewall   F. 点火器插头
7. hydraulic accumulator   G. 增压泵
8. igniter plug   H. 电磁活门
9. ignition exciter   I. 燃油活门
10. boost pump   J. 防火墙

Section 3 Aviation Translation

Translate the following English sentences into Chinese.

1. An auxiliary power unit (APU) is a small, multipurpose gas turbine engine.

2. The APU generator supplies the electrical power to the airplane electrical system.

3. The APU five warning light, the APU fire warning horn, the APU fire bottle discharge switch, and the APU remote shut down switch are on this panel.

4. The breaker has six main 115 volt AC contacts and 20 auxiliary contacts.

5. The closing and tripping signals for the breakers are from the P5-4 panel.

Section 4 Aviation Writing

Situations: Yafeng Guo is an aircraft mechanic. He is inspecting the plane and finding some faults. Please help him write down fault description. Some hints of the description words, phrases & terms and the key sentences about fault location, fault description, fault solutions are offered as follows.

Key words, phrases & terms:

1. APU 进气口   air inlet for APU

2. APU 燃油活门　　　　　　fuel valve for the APU

3. APU 供油　　　　　　　　APU feed

4. APU 勤务灯　　　　　　　APU service light

5. APU 维护灯　　　　　　　APU MAINT light

6. APU 超速灯　　　　　　　APU OVER SPEED light

7. APU 引气活门　　　　　　APU bleed air valve

8. APU EGT 指示器　　　　　APU EGT indicator

9. APU 探测不工作灯　　　　APU DET INOP light

10. APU 灭火瓶释放灯　　　　APU BOTTLE DISCHARGED light

11. APU 滑油压力低灯　　　　APU LOW OIL PRESSURE light

12. APU 自测不工作信息　　　APU BITE INOP message

13. APU 火警探测器故障　　　APU fire detector fault

14. APU 电瓶　　　　　　　　APU battery

15. APU 控制器　　　　　　　APU controller (APUC)

Key sentences:

1. APU/ 发动机左 / 右灭火瓶已释放。APU/Engine left/right bottle discharged.

2. 主发起动时 APU 排气口瞬时出现火焰。

    Momentary flames from APU exhaust during main engine start.

3. 操作测试 APU 火警探测系统。Do the operational test of the APU fire detection system.

4. 在摇控 APU 灭火控制面板上作动时，不释放灭火剂。

    Does not release extinguishant when activated at the remote APU fire control panel.

5. 尾喷管冒火焰。Backfire.

6. 没有起动，故障灯亮。APU can not be started, FAULT light was on.

7. 没有起动，无 APU 指示灯亮。APU can not be started, no APU indication lights were on.

8. 没有起动，超速灯亮。APU can not be started, OVER SPEED light was on.

9. 火警假警告。Fire false alarm.

10. 燃油活门不全开 / 全关。Fuel valve did not open/ close fully.

11. 尾气烟雾 / 火焰。Smoke or flame in the exhaust.

12. 主发起动时，APU 排气出现刺耳噪音。

    APU exhaust made sharp noise when main engine started.

Task: Yafeng Guo finds out three problems in the check, the first, the air inlet door could not open fully, the second, APU could not be started, and maintenance light was on, the last, the light was on when the APU generator was on line.

Lesson 11 Wings

Learning Objectives:

1. Knowledge objectives:

   A. To master the major words, related terms and abbreviations about wings.

   B. To master the key sentences.

   C. To understand the different control surfaces on the wings.

2. Competence objectives:

   A. To be able to understand frequently-used & complex sentence patterns, acronyms and obtain key information on aviation maintenance quickly.

   B. To be able to talk about aviation or aircraft in English.

   C. To be able to fill in job cards in English.

3. Quality objectives:

   To be able to self-study with the help of aviation dictionaries, Internet or other resources.

Section 1 Aviation Listening and Speaking

1.1 Aviation Listening: listen to the audio and fill in the blanks with the missing words.

听力录音

Very often you can see the tail 1._____ of an Airbus pointing into the right direction once parked at the gate and shut down 2. _____. The ailerons and rudders are primarily powered by three independent 3. _____ systems, which provide 4._____ to the actuators. Once you shut down the engines the hydraulic pressure drops 5. _____ the pressure is generated by a hydraulic gearbox connected to the 6. _____ engines. Once the pressure is lost, the weight of the 7. _____ press the cylinders of the actuators inward, making the ailerons look like as if they are 8._____ down. The massive tail rudder is also 9. _____and turns either left or right, depending on the 10. _____ direction.

1.2 Aviation Speaking: look at the pictures below and describe them in details.

Clues: wings; flaps; slats; spoilers.

Section 2 Aviation Reading

课文朗读录音   译文

Pre-reading questions:

1. What are the functions of the wings?
2. What's the basic structure of the wings?

Wings

Introduction

All the components are designed to fulfill some particular functions. The first function of the wings is to provide the principal lifting force of an airplane. Secondly, they are controlled by pilots to help an aircraft to change its movement direction. Lastly, for most of aircraft, wings are the parts to locate engines and landing gears.

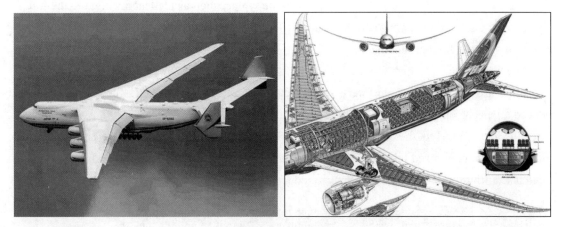

Fig.11-1   Wings

The wing provides the principal lifting force of an airplane. Lift is obtained from the dynamic action of the wing with respect to the air. The cross-sectional shape of the wing as viewed from the side is known as the airfoil section. The planform shape of the wing (the shape of the wing as viewed from above) and placement of the wing on the fuselage (including the angle of incidence), as well as the airfoil section shape, depend upon the airplane mission and the best compromise necessary in the overall airplane design.

Basic Structure

微课

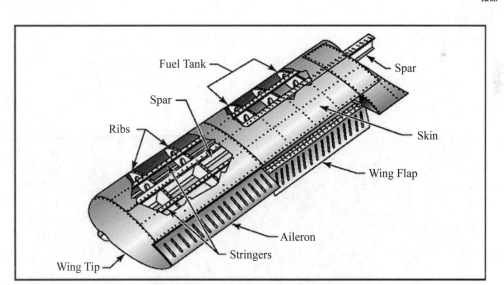

Fig. 11-2   Main Components of a Wing Structure

The graph above shows the main components of a wing structure. Firstly, a wing structure has the front spar & rear spar, ribs, upper & lower skins and stringers to form the frame and body. The internal space of the wing frame is for the fuel tank and electrical system. Basic wing structure consists of left, center, and right wing sections.

Various Control Surfaces

There are various control surfaces located on the wings, such as the leading edge & the trailing edge flaps, the slats, the ailerons and the spoilers. They are hydraulically powered and only extended or retracted during take-off and landing.

The flaps

Flaps are hinged or pivoted parts of the leading and/or trailing edges for increasing lift at reduced airspeeds during landing and take-off. They are designed for reducing flying speed as well. When the flaps deployed, the curvature of the wing is increased. The speed of the aircraft will reduce but lift force is maximized to make sure the aircraft can fly safely under lower speed. Canoe-shaped flap track fairings are installed under flaps to reduce drag.

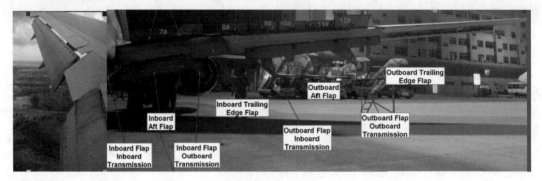

Fig. 11-3   The Flaps

The ailerons

Ailerons are the hinged parts on the outbound trailing edge for rolling the wings from side to side. They are used to generate a rolling motion for an aircraft. By controlling one aileron to move up and the other to move down, lift forces on the two wings are unbalanced and lead the aircraft change its path.

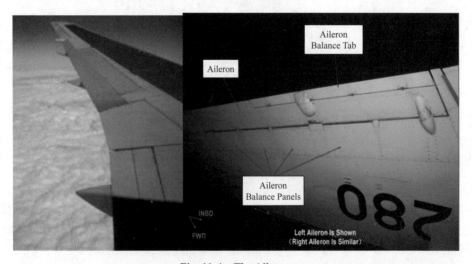

Fig. 11-4   The Aileron

The spoilers

Spoilers are used to disrupt the airflow so as to reduce the lift on an airplane wing quickly. By operating independently on each wing, they may provide an alternate form of roll control. Spoilers are located on the top of the upper surface of wings. By rising up, the spoilers generate a controlled stall. As a result, the lift force will reduce sufficiently.

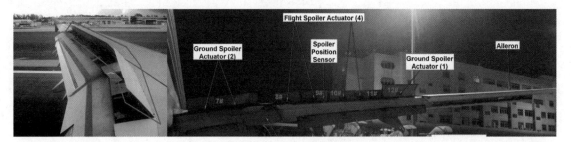

Fig. 11-5　The Spoilers

The slats

Slats at the front part of the wing are used during take-off and landing to produce additional lift. For fixed-wing aircraft, leading edge slats are installed on the leading edge of a wing. They are designed into streamline shape to reduce drag. During takeoff and landing, the slats can be used to change the attack angle of wing to produce an additional force.

Fig. 11-6　The Slats

Wings-Top View

Fig.11-7 shows the left wing. The right wing is almost the same as the left wing. Most of material in the wing is aluminum. These components attach to the wing structure:

* Engine nacelle/pylon
* Flight control surfaces
* Wingtips
* Winglets

Access panels

Access panels on the top of the wing gives access the main landing gear support structure.

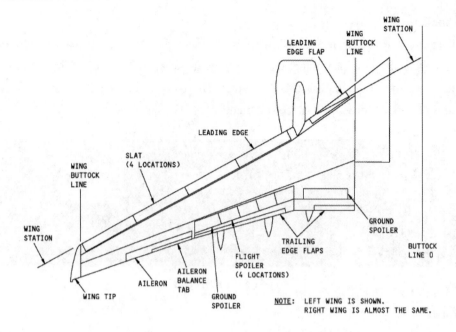

Fig.11-7 Wings-Top View

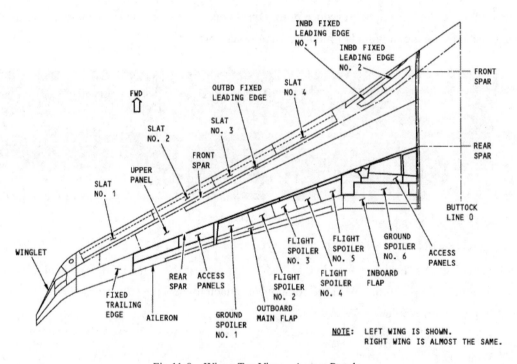

Fig.11-8 Wings-Top View—Access Panels

Wingtips

The wingtip has anti-collision lights and forward and aft position lights. Three access panels on the bottom of the winglet gives access to the electrical connectors from the wing and the electrical connectors inside the winglet for the anti-collision lights and forward position lights.

Winglets

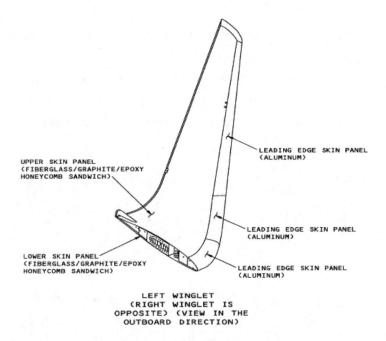

UPPER SKIN PANEL
(FIBERGLASS/GRAPHITE/EPOXY
HONEYCOMB SANDWICH)

LEADING EDGE SKIN PANEL
(ALUMINUM)

LOWER SKIN PANEL
(FIBERGLASS/GRAPHITE/EPOXY
HONEYCOMB SANDWICH)

LEADING EDGE SKIN PANEL
(ALUMINUM)

LEADING EDGE SKIN PANEL
(ALUMINUM)

LEFT WINGLET
(RIGHT WINGLET IS
OPPOSITE) (VIEW IN THE
OUTBOARD DIRECTION)

Fig.11-9 Winglet

Winglets are designed to make the flight more efficient. They are the extension of wings edge. They can increase the lift force and reduce the lift-induced drag force efficiently. Generally, winglets are designed into V-shape.

The upper and lower skin panels of the winglets are made of composite fiberglass/graphite/epoxy honeycomb material with an aluminum leading edge and aluminum ribs. The winglet has anti-collision lights and forward and aft position lights. An access panel on the bottom of the winglet gives access to the electrical connectors from the wing and the electrical connectors inside the winglet for the anti-collision lights and forward position lights. Access for the aft position light is in the winglet aft canoe.

Older winglets have a single lens and a single mount bracket for the forward position lights and the anti-collision lights.

Newer winglets have two lens and two mount brackets for separate forward position lights and a separate anti-collision light in each winglet.

Weight added to the winglet can cause a safety condition. The old paint must be removed before new paint is applied to the winglet. No new paint or logos maybe applied over the old paint. You must get the approval of Aviation Partners—Boeing, LLC for any changes.

Wings-Bottom View

The bottom of the left wing is shown. The right wing is almost the same.

Fuel tank (or wet bay) access panels allow access into the center fuel tank, the main fuel tanks and the surge tanks.

- Dry bay access panels allow access into the dry bay tanks which are outboard of the surge tanks.
- Access panels behind the wing leading edge gives access to the forward spar structure and the leading edge slat actuators and tracks.
- Access panels aft of the rear spar gives access to the rear spar structure, fuel system pumps, aileron and the trailing edge flaps mechanisms.
- Access panels in the wing tip gives access to the anti-collision lights and position lights.

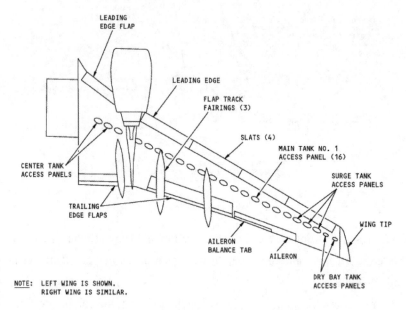

Fig. 11-10　Wings-Bottom View

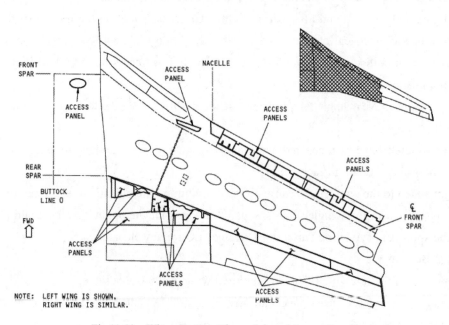

Fig.11-11　Wings-Bottom View—Inboard Access Panels

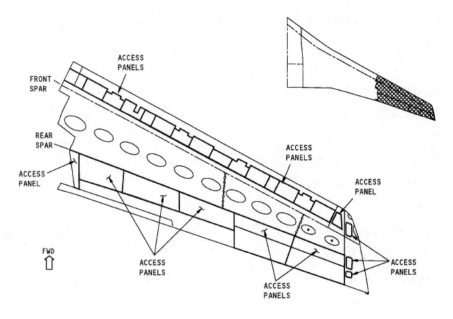

Fig.11-12   Wings-Bottom View—Outboard Access Panels

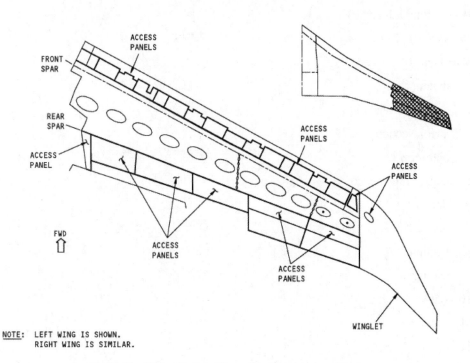

Fig.11-13   Wings-Bottom View—Outboard Access Panels

Abbreviations and Acronyms

1.   Control (CNTL)   控制，操纵

2.   Direct current (DC)   直流电

3.	Flight Control Computer (FCC)	飞行控制计算机
4.	Flight (FLT)	飞行
5.	Ground (GND)	地面
6.	Leading Edge (LE)	前缘
7.	Main Landing Gear (MLG)	主起落架
8.	Nose Landing Gear (NLG)	前起落架
9.	Power Control Unit (PCU)	电源控制组件
10.	Pounds per Square Inch (PSI)	磅／平方英寸
11.	Rudder (RUD)	方向舵
12.	Speed Brake (S/B)	减速板
13.	Section (SEC)	截面，区
14.	Shutoff Valve (SOV)	关断活门
15.	Stabilizer (STAB)	安定面
16.	Standby (STBY)	备用
17.	Switch (SW)	开关，电门
18.	System (SYS)	系统
19.	Trailing Edge (TE)	后缘
20.	Typical (TYP)	典型的

Words and Phrases

1.	dynamic *adj.*	动力的
2.	cross-sectional *adj.*	横截面的
3.	airfoil section	翼剖面
4.	planform *n.*	平面图；俯视图
5.	angle of incidence	入射角
6.	flap *n.*	襟翼
7.	pivoted *adj.*	转动的，回转的
8.	leading *adj.*	前面的
9.	spoiler *n.*	扰流板
10.	slat *n.*	缝翼
11.	nacelle *n.*	吊舱；短舱
12.	pylon *n.*	吊架
13.	winglet *n.*	翼梢小翼
14.	wingtip *n.*	翼尖
15.	anti-collision light	防撞灯
16.	epoxy *n.*	环氧基树脂

17. epoxy honeycomb	环氧蜂窝
18. single lens	单透镜
19. surge tank	通风油箱；浪涌油箱
20. actuator *n.*	作动筒；制动器
21. track *n.*	轨条；滑轨
22. pump *n.*	泵
23. rib *n.*	翼肋
24. surge effect	浪涌效应；波涌效应
25. Krueger flap	克鲁格襟翼
26. roll control	侧滚控制；横滚控制
27. airspeed *n.*	空速

Exercises

2.1 Answer the following questions according to the passage.

1. What are the functions of the wing?

2. What are the main components of a wing structure?

3. What control surfaces are located on the wings?

4. What are the functions of three access panels on the bottom of the winglet?

5. What are the functions of winglets?

2.2 Translate the following terms or abbreviations into Chinese.

1. dynamic   6. flap

2. cross-sectional   7. slat

3. airfoil section   8. aileron

4. planform   9. spoiler

5. angle of incidence   10. nacelle

2.3 Give the equivalent English terms and corresponding Chinese translations according to the pictures.

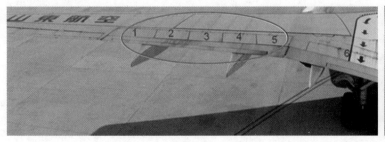

1._____   2. _____

3._____

4._____

5._____

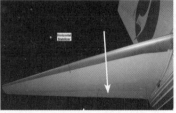

6._____

Section 3 Aviation Translation

Translate the following English sentences into Chinese.

1. Lift is obtained from the dynamic action of the wing with respect to the air.

2. Three access panels on the bottom of the winglet gives access to the electrical connectors from the wing and the electrical connectors inside the winglet for the anti-collision lights and forward position lights.

3. The upper and lower skin panels of the winglets are made of graphite /graphite/epoxy honeycomb material with an aluminum leading edge and aluminum ribs.

4. There are various control surfaces located on the wings,such as the leading edge & the trailing edge flaps, the slats, the ailerons and the spoilers.

5. Weight added to the winglet can cause a safety condition.

Section 4 Aviation Writing

Situations: Yafeng Guo is an aircraft mechanic. He is inspecting the plane and finding some faults. Please help him write down fault description. Some hints of the description words, phrases & terms and the key sentences about fault location, fault description, fault solutions are offered as follows.

Key words, phrases & terms:

1. 机翼　　　　　　　　　　　　wing
2. 前缘　　　　　　　　　　　　leading edge/LE
3. 后缘　　　　　　　　　　　　trailing edge/TE
4. 接近盖板　　　　　　　　　　access panel

5. 止动块   stop pad

6. 翼身整流罩   wing to body fairing

Key sentences:

1. 右大翼（机翼前缘位置）位置灯灯罩距翼尖 9cm 处灯罩有一处 2.7cm 的裂纹。

   There is a 2.7cm-long crack at the forward right wing position cover, which is 9cm away from wing tip.

2. 进行前缘襟翼主滑轨止动块组件检查时，机翼固定前缘下翼面的接近盖板有多个螺钉无法取下。

   Several bolts of the access panel below the leading edge could not be removed while inspecting the stop ASSY of the leading flap's main track.

3. 左大翼前缘 4 号缝翼后一组止动块丢失。

   A group of stop pads after the L/H wing No.4 slot is missing.

4. 在执行翼身过热探测系统普查中发现 M1762 阻值偏低，只有 0.5m 欧姆。

   The resistance of the M1762 is too low which is only 0.5m in the wing in body overheats detecting system survey.

Task: Yafeng Guo finds out three problems in the check. The first, there is a bird impact at the right leading edge at P46 check, the second, he finds out thunder striking points on the lower surface of aileron tab, the last, it is difficult to remove several screws for the right wing leaks fuel. Please finish the writing tasks for him.

Lesson 12 Flight Control Surfaces

Learning Objectives:

1. Knowledge objectives:

A. To master the major words, related terms and abbreviations about flight control surfaces.

B. To master the key sentences.

C. To know the functions of different flight control surfaces.

2. Competence objectives:

A. To be able to understand frequently-used & complex sentence patterns, acronyms and obtain key information on aviation maintenance quickly.

B. To be able to talk about aviation or aircraft with English.

C. To be able to fill in job cards in English.

3. Quality objectives:

To be able to self-study with the help of aviation dictionaries, Internet or other resources.

Section 1 Aviation Listening and Speaking

1.1 Aviation Listening: listen to the audio and fill in the blanks with the missing words.

听力录音

You most certainly have seen it on your last flight with an 1._____ A320. We'll talk about a small little triangle which you can find within the 2._____, indicating the wings leading edge and 3._____ edge, but why is that important? The main reason for the triangle is, to help the pilot or a flight 4._____ find the leading/trailing edge quicker pilot 5._____ there is a problem with the flaps or 6._____. Especially during night it will make it easier to find the 7._____ to look out of. But see more about that in the video.

The 8._____ mysterious object I'll be talking about is the little yellow hook 9._____ on the Airbus wing. I've heard many funny rumors about the hook and it's purpose. It's all about the Airbus escape ropes in case of an 10._____.

1.2 Aviation Speaking: look at the pictures below and describe them in details.

Clues: flight control surface; flap lever; Flap-slat Electronics Unit; leading edge; trailing edge.

Section 2 Aviation Reading

Pre-reading questions:

1. What are the functions of the flight control?
2. How many ways is flight control divided into?

课文朗读录音   译文

Flight Control Surfaces

The flight controls keep the airplane at the necessary attitude during flight and provides maneuvering control about the lateral, longitudinal, and vertical axes. They also provide increased lift for taking off and landing as well as increased aerodynamic drag both in flight and on ground.

The control surfaces include all those moving surfaces of an airplane used for attitude, lift, and drag control on the wing and the empennage. They include the tail assembly, the structures at the rear of the airplane that serve to control and maneuver the aircraft and structures forming part of and attached to the wing. They are hinged moveable surfaces on the wings, horizontal stabilizer and vertical stabilizer for airplane control during flight and high speed ground operations.

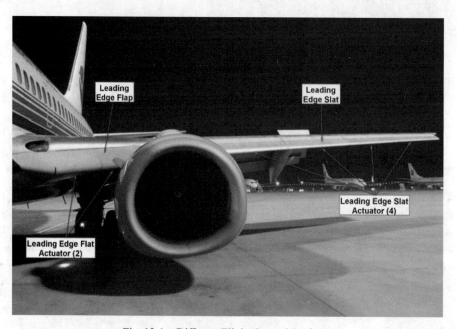

微课

Fig. 12-1　Different Flight Control Surfaces

HYDRAULIC ACTUATORS OR ELECTRIC MOTOR MOVE THE SURFACES, SO THE PERSONNEL MUST BE VERY CAREFUL WHEN WORKING NEAR FLIGHT CONTROL SURFACES. WHEN HYDRAULIC POWER IS ON, MAKE SURE THAT ALL THE FLIGHT CONTROL SURFACES ARE CLEAR OF PERSONNEL AND EQUIPMENT.

Primary Flight Controls

Flight controls are divided into primary controls and secondary controls. The primary flight controls consist of the ailerons, elevators and rudder. The secondary flight controls consist of the spoilers, trailing edge flaps, leading edge devices and the horizontal stabilizer. Normal operation of the primary controls is hydraulic power supplied by system A & B. Either hydraulic system operating alone can provide effective control of the primary flight control.

Let's take the example of Boeing 737.

The subsystems of the primary flight controls are as follows.

➢　Aileron (2)

➢　Elevator (2)

➢　Rudder

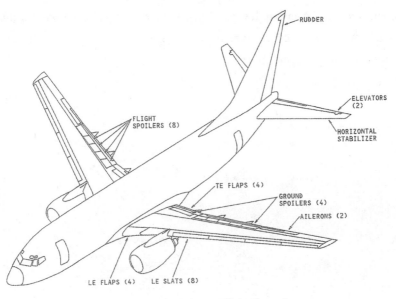

Fig.12-2   The Primary Flight Controls

Secondary Flight Controls

The subsystems of the secondary flight controls are as follows.

➢   leading edge flaps (4)

➢   leading edge slats (8)

➢   trailing edge flaps (4)

➢   spoilers (12)

➢   horizontal stabilizer

The 12 spoilers numbered 1 to 12 from the left to the right consist of 4 ground and 8 flight spoilers. The inboard flight spoilers are powered by hydraulic system A while outboard flight spoilers are powered by hydraulic system B.

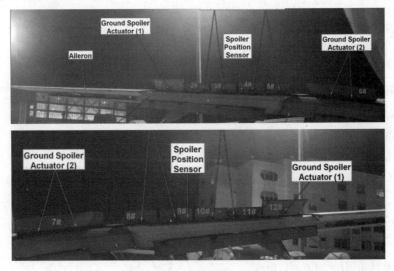

Fig. 12-3   Spoiler No.1-No.12

Ground spoilers are two position devices: full down or full up. All ground spoilers are powered by hydraulic system A through a control valve and a bypass valve. Flight spoilers function as speed brakes both in flight and on the ground.

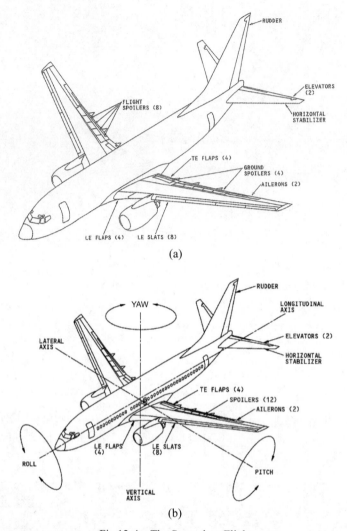

(a)

(b)

Fig.12-4　The Secondary Flight

Control Surfaces on the Tail

The tail section includes the tail assembly, the structures at the rear of the airplane that serve to control and maneuver the aircraft.

The tail usually has a fixed horizontal piece (called the horizontal stabilizer) and a fixed vertical piece (called the vertical stabilizer). The stabilizers provide stability for the aircraft—they keep it flying straight. The vertical stabilizer keeps the nose of the plane from swinging from side to side (called yaw), while the horizontal stabilizer prevents an up-and-down motion of the nose (called pitch).

Horizontal stabilizer and elevator

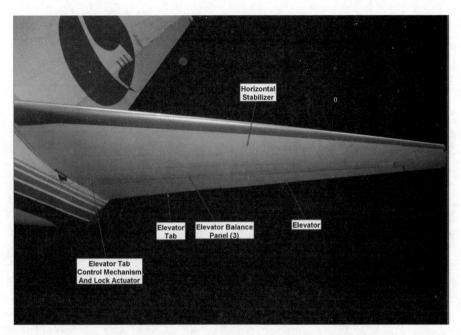

Fig. 12-5　Control Surfaces on the Horizontal Stabilizer

The stabilizer trim wheels move cables that give an input to the gearbox. The gearbox moves a jackscrew and moves the stabilizer. The electric stabilizer trim switches control an electric motor near the gearbox. The motor moves the gears to move the stabilizer. The autopilot also controls the stabilizer trim motor. When the stabilizer moves, it also moves the elevators through the elevator feel and centering unit.

Elevator is a hinged moveable part on the trailing edge of the horizontal stabilizer for airplane pitch (nose up and down) control. Moving the elevators out of their faired position causes the airplane to change the pitch altitude about the lateral axis. The amount of elevator movement establishes the airplane pitch rate. When the desired altitude is reached, the elevators are faired and the air-plane maintains the new pitch altitude.

The control columns move cables that give input to the elevator feel and centering unit. This controls the elevator PCUs (power control unit). The PCUs move torque tubes that move the elevators. The autopilot actuators give a mechanical input to the PCUs through the feel and centering unit. The PCU moves the elevators.

Vertical stabilizer and rudder

Rudder is a hinged moveable part on the trailing edge of the vertical stabilizer for airplane yaw (nose to the left or right).

The rudder is mounted on the after side of the vertical stabilizer. Control about the vertical axis is provided by the rudder. Control is provided by a single rudder without tab. The rudder is pedal operated by the captain or the first officer. Pedal movement rotates the forward quadrants, which are cable connected to the aft quadrant. Rotation of the aft quadrant moves a control rod connected to a torque tube. Rotation of the torque tube moves a crank connected to the rudder power control unit linkage. This admits hydraulic fluid to the actuating cylinder, which moves the rudder.

The rudder pedals move cables that give an input to the rudder feel and centering unit. This controls the rudder PCUs. The rudder PCUs move the rudder. The rudder trim switch gives an input to the rudder feel and centering unit and change the neutral position of the rudder.

Fig. 12-6   Control Surfaces on the Vertical Stabilizer

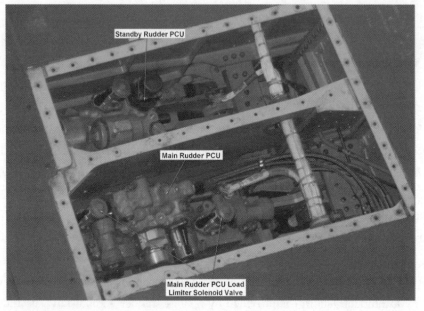

Fig. 12-7   Rudder PCU

Trim tabs

Trim tabs are small moving sections attached by hinges at the rear of the aileron surfaces, elevators and rudders. The function is to balance the airplane to fly in a stable cruise condition when it is too nose heavy, tail heavy, or wing heavy. And it can also maintain the elevator, rudder and ailerons at whatever setting the pilot wishes without the pilot maintaining pressure on the controls. Lastly it helps move the elevators, rudder, and ailerons and thus relieves the pilot of the effort necessary to move the surfaces.

Control Surfaces on the Wing

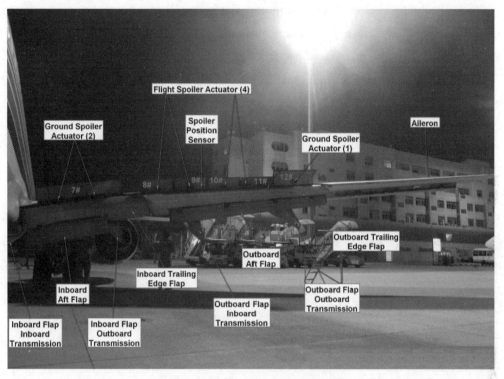

Fig. 12-8   Control Surfaces on the Wing

Aileron

The aileron is a hinged moveable part on the trailing edge of each wing for airplane roll control. They are located on the outboard trailing edge of each wing.

The aileron control wheels move cables that give input to the aileron feel and centering unit. This controls the aileron PCUs. The PCUs move the aileron wing cables and the ailerons. The aileron trim switches give an input to the aileron feel and centering unit and change the neutral position of the ailerons.

Aileron PCU movement also goes to the spoiler mixer. The mixer moves the flight spoiler wing cables, which control the flight spoiler actuators. The actuators move the flight spoilers to assist the ailerons for roll control.

The autopilot actuators give a mechanical input to the PCUs through the feel and centering unit. The PCUs move the aileron wing cables and the ailerons and they give input to the spoiler mixer.

The aileron control system provides the means of rotating the airplane about the longitudinal axis. The aileron control system is actuated by rotation of either the captain's or first officer's aileron control wheel. The autopilot system also provides a control input. These inputs control a hydraulic powered system which drives the ailerons. Moving the ailerons out of their trimmed position causes the airplane to rotate and change altitude.

Flaps and slats

The flap control lever moves the trailing edge flap control valve. Hydraulic pressure goes through the valve and drives the hydraulic motor. The hydraulic motor supplies power to the flap drive system and the flaps move. Follow-up cables give feedback to the trailing edge flap control valve to stop the flaps at the commanded position.

The follow-up cables also give an input to the leading edge flaps control valve. This controls the position of the leading edge devices. Then hydraulic pressure goes to the actuators and moves the leading edge flaps and slats.

The alternate flap switches electrically control the trailing edge flaps. They also control the standby hydraulic system to extend the leading edge flaps and slats.

Spoilers and speed brakes

The speed brake lever moves cables that control the spoiler mixer. The mixer moves the spoiler wing cables that control the flight spoiler actuators. The mixer also moves the ground spoiler control valve. The valve supplies hydraulic pressure to the ground spoiler actuators to raise the spoilers. During automatic deployment, the auto speed brake actuator gives input to the same cables as above and back drives the speed brake lever.

Abbreviations and Acronyms

1. Control Wheel (C/W)　　　　　　控制盘
2. Circuit Breaker (CB)　　　　　　跳开关
3. Control (CNTL)　　　　　　　　控制
4. Flight Control Computer (FCC)　飞行控制计算机
5. Flight (FLT)　　　　　　　　　　飞行
6. Ground (GND)　　　　　　　　　地面
7. Leading Edge (LE)　　　　　　　前缘
8. Main Landing Gear (MLG)　　　　主起落架
9. Nose Landing Gear (NLG)　　　　前起落架
10. Power Control Unit (PCU)　　　　电源控制组件
11. Pound per Square Inch (PSI)　　　磅 / 平方英寸
12. Rudder (RUD)　　　　　　　　　方向舵
13. Speedbrake (S/B)　　　　　　　　减速板
14. Section (SEC)　　　　　　　　　截面 , 区
15. Shutoff Valve (SOV)　　　　　　关断活门

16. Stabilizer (STAB)	安定面
17. Standby (STBY)	备用
18. System (SYS)	系统
19. Trailing Edge (TE)	后缘

Words and Phrases

1. surface *n.*	舵面
2. flight control	飞行操纵
3. maneuvering control	机动控制
4. longitudinal *adj.*	纵向的
5. axes *n.*	(axis 的复数形式) 坐标轴
6. vertical axis	垂直轴
7. lift *n.*	升力
8. aerodynamic *adj.*	空气动力的
9. hinged *adj.*	有（用）铰链的
10. horizontal stabilizer	水平安定面
11. vertical stabilizer	垂直安定面
12. actuator *n.*	作动筒，传动装置
13. aileron *n.*	副翼
14. elevator *n.*	升降舵
15. rudder *n.*	方向舵
16. trailing edge flap	后缘襟翼
17. leading edge device	前缘设备
18. device *n.*	装置
19. maneuver *v.*	调动，机动
20. attach to	与……相连接
21. stability *n.*	稳定性
22. straight *adj.*	直线的
23. yaw *n.*	偏航
24. pitch *n.*	俯仰
25. cable *n.*	钢索
26. gearbox *n.*	变速箱
27. jackscrew *n.*	致动螺杆
28. center of lift (CL)	升力中心
29. center of gravity (CG)	重力中心
30. faired position	中立位
31. axis *n.*	轴

32. control column	操纵杆
33. torque tube	扭力管；转矩管
34. input *n./v.*	输入
35. pedal *n.*	脚蹬；踏板
36. quadrant *n.*	扇形盘 / 摇臂
37. crank *n.*	曲柄
38. actuating cylinder	作动筒
39. trim tab	配平调整片
40. rotate *v.*	转动；旋转
41. actuate *v.*	作动；驱动
42. rotation *n.*	转动；旋转
43. attitude *n.*	姿态
44. hydraulic *adj.*	液压的
45. follow-up cable	跟随钢索
46. feedback *n.*	反馈
47. speed brake	减速板
48. deployment *n.*	展开
49. back drive	反向传动；反向驱动

答案

Exercises

2.1 Answer the following questions according to the passage.

1. How many control surfaces does an airplane have?

2. How many spoilers does an airplane have?

3. What is the function of the trim tab?

4. Where are the ailerons located?

5. What is the function of the flap control lever?

2.2 Translate the following terms or abbreviations into Chinese.

1. lift	2. back drive
3. actuate	4. follow-up cable
5. rotate	6. gearbox
7. speed brake	8. control column
9. maneuver	10. maneuvering control

2.3 Give the equivalent English terms and corresponding Chinese translations according to the pictures.

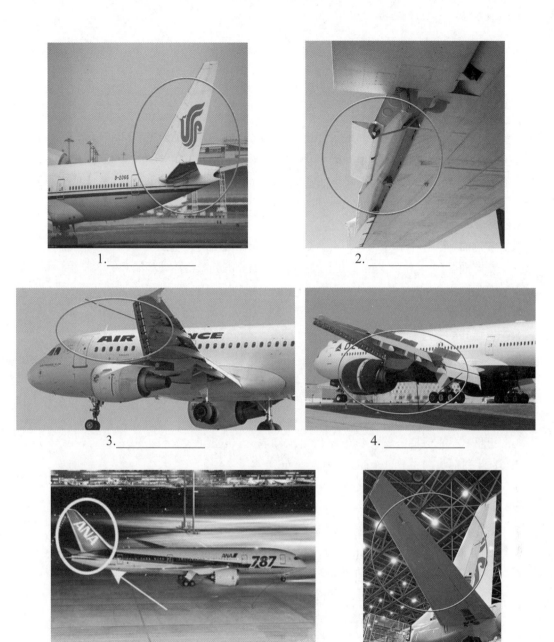

1._____

2. _____

3._____

4. _____

5. _____

6. _____

Section 3 Aviation Translation

Translate the following English sentences into Chinese.

1. The flight controls keep the airplane at the necessary attitude during flight.

2. The inboard flight spoilers are powered by hydraulic system A while outboard flight spoilers are powered by hydraulic system B.

3. The stabilizers provide stability for the aircraft—they keep it flying straight.

4. The vertical stabilizer keeps the nose of the plane from swinging from side to side (yaw), while the horizontal stabilizer prevents an up-and-down motion of the nose (pitch).

5. When hydraulic power is on, make sure that all the flight control surfaces are clear of personnel and equipment.

Section 4 Aviation Writing

Situations: Yafeng Guo is an aircraft mechanic. He is inspecting the plane and finding some faults. Please help him write down fault description. Some hints of the description words, phrases & terms and the key sentences about fault location, fault description, fault solutions are offered as follows.

Key words, phrases & terms:

1. 安定面   stabilizer
2. 安定面配平   stabilizer trim
3. 人工配平手轮   stabilizer trim wheel
4. 安定面配平电门   stabilizer trim switch
5. 低速移动   move with slow speed
6. 机头向上   nose up
7. 机头向下   nose down
8. 安定面失去配平指示灯   STAB OUT OF TRIM light

Key sentences:

1. 飞行中不能保持位置。Cannot hold position in flight.
2. 非计划机头向上。Unscheduled nose up.
3. 自动驾驶接通 CMD A/B。Autopilot engaged in CMD A/B.
4. 配平轮惯性转动超过……Wheel coasted more than...
5. 安定面配平电门在向上配平时不工作。

   Stabilizer trim switches cannot work during triming up.
6. 机身左侧平尾处有液压油。

   The L/H horizontal stabilizer of the fuselage has hydraulic fluid.
7. 润滑安定面螺杆。Lubricate stabilizer jackscrew.
8. 功能测试次级安定面配平制动器。

   Do a functional test of secondary stabilizer trim brake.

Task: Yafeng Guo finds out three problems in the check. The first, the stabilizer cannot move freely, the second, it can only move with slow speed, and the last, stabilizer trim switches cannot work when triming up. Please finish the writing tasks for him.

Appendix I Comparison of ATA 100 Chapters in English & Chinese

CHAPTER 章	SUBJECT	主题
5	**TIME LIMITS/MAINT.CHECK**	**时限 / 维护检查**
6	**DIMENSION & AREAS**	**尺寸及区域划分**
10	Fuselage	机身
20	Wing-Station & Access	机翼—站位及盖板
30	Nacelle/Pylon-Station & Access	短舱 / 吊架—站位及盖板
40	Tail Assembly-Stations & Access	尾翼—站位及盖板
7	**LIFTING & SHORING**	**顶起和支撑**
10	Jacking	顶起
20	Shoring	支撑
8	**LEVELING & WEIGHING**	**校水平和称重**
10	Leveling	校水平
20	Weighing Procedures	称重程序
30	Weight & Balance Data	重量平衡数据
9	**TOWING & TAXIING**	**牵引和滑行**
10	Towing	牵引
20	Hand Signaling	手势
30	Taxiing	滑行
10	**PARKING & MOORING**	**停放和系留**
10	Parking	停放
20	Mooring	系留
11	**PLACARDS & MARKINGS**	**铭牌及标志**
10	Exterior Color Schemes & Markings	外部颜色设计和标志
20	Exterior Placards & Markings	外部铭牌和标志
30	Interior Placards & Markings	内部铭牌和标志
12	**SERVICING**	**勤务工作**
10	Refueling-Defueling	加油 / 放油
20	Oil Service	滑油勤务
30	Hydraulic Servicing	液压系统勤务

CHAPTER 章	SUBJECT	主题
40	Water Service	水系统勤务
50	Waste Service	废物处理勤务
60	Gaseous Service	补充气体勤务
70	TYRE Service	轮胎勤务
75	Fire Extinguisher	灭火瓶
80	Rain Repellent	排雨
90	Lubrication	润滑
14	**EQUIPMENT OPERATION**	**设备操作**
10	Ground Operation Certification	地面操作证书
20	Engine Operation/System Check (Run up)	发动机操作 / 系统检查（试车）
30	Discrepancy Analysis & Correction (Troubleshooting)	故障分析及纠正（排故）
15	**TRAINING OUTLINE**	**训练大纲**
16	**GROUND SUPPORT EQUIPMENT**	**地面支援设备**
10	Fixed Refuel Equipment	固定加油设备
20	Motorized Vehicles	机动车辆
30	Towed Vehicles	牵引车辆
40	External Ground Power	辅助地面电源
17	**FACILITIES EQUIPMENT**	**设施与设备**
10	Heating System	加温系统
20	Air Conditioner	空调
30	Compressed Air	压缩空气
40	Air Handling	通风
50	Emergency Power	应急电源
60	Fire Protection	防火
70	Servicing	勤务
80	Lifting Equipment	起重设备
90	Miscellaneous Equipment	其他设备
20	**STANDARD PRACTICES-AIRFRAME**	**标准工艺—机身**
21	**AIR CONDITIONING**	**空调系统**
10	Compression	压气机
20	Distribution	分配
30	Pressurization Control	增压控制
40	Heating	加温
50	Cooling	冷却

CHAPTER 章	SUBJECT	主题
60	Temperature Control	温度控制
70	Moisture/Air Contaminant Control	湿度 / 空气污染控制
22	**AUTO FLIGHT**	**自动飞行**
10	Autopilot	自动驾驶
20	Speed-Attitude Correction (Mach Trim)	速度—姿态修正（马赫配平）
30	Auto Throttle	自动油门
40	System Monitor	系统监控
50	Aerodynamic Load Alleviating	减压气动力负荷装置
23	**COMMUNICATIONS**	**通信**
10	HF	高频
20	VHF	甚高频
30	Passenger Address & Entertainment	旅客广播和娱乐系统
40	Interphone	内话
50	Audio Integrating	音频综合系统
60	Static Discharging	放静电
70	Audio Monitoring	音频监听（舱内录音机）
80	Integrated Automatic Tuning	综合自动调谐
24	**ELECTRICAL POWER**	**电源**
10	Constant Speed Drive (CSD)	恒速传动装置
20	AC Generation	交流发电
40	External Power	外部电源
50	Electrical Load Distribution	电负载分配
25	**EQUIPMENT/FURNISHINGS**	**机舱设备 / 装饰**
10	Flight Compartment	驾驶舱
20	Passenger Compartment	客舱
30	Buffet and Galley	餐具间和厨房
40	Lavatories	盥洗室
50	Cargo Compartments	货舱
60	Emergency	应急设备
70	Accessory Compartments	附件舱
80	Insulation	绝缘材料
26	**FIRE PROTECTION**	**防火系统**
10	Detection	探测
20	Extinguishing	灭火

CHAPTER 章	SUBJECT	主题
30	Explosion Suppression	防爆
27	**FLIGHT CONTROLS**	**飞行操纵**
10	Aileron Tab Control	副翼调整片操纵
20	Rudder and Tab Control	方向舵和调整片操纵
30	Elevator and Tab Control	升降舵和调整片操纵
40	Horizontal Stab Control	水平安定面操纵
50	Flaps-Trailing Edge Control	后缘襟翼操纵
60	Spoiler, Drag Devices and Variable Aerodynamic Fairings Control	扰流板，减速装置和可变气动整流操纵
70	Gust Lock and Dampener Control	阵风锁定及阻尼控制
80	Lift Augmenting Control	增升装置操纵
28	**FUEL**	**燃油**
10	Storage	油箱
20	Distribution	分配
30	Dump	放油
40	Indicating (Not Engine)	指示（不包括发动机）
29	**HYDRAULIC POWER**	**液压**
10	Main	主系统
20	Auxiliary	辅助系统
30	Indicating	指示
30	**ICE & RAIN PROTECTION**	**防冰和排雨**
10	Airfoil	翼面
20	Air Intakes	进气口
30	Pitot and Static	动静压
40	Windows, Windshields & Doors	窗，风挡和门
50	Antennas and Radomes	天线和雷达罩
60	Propellers/Rotors	螺旋桨/旋翼
70	Water lines	水管
80	Detection	探测
31	**INDICATING/RECORDING SYSTEM**	**指示/记录系统指示/记录系统**
10	Panels	仪表板
20	Independent Instruments	独立仪表
30	Recorders	记录器
40	Central Computers	中央计算机

CHAPTER 章	SUBJECT	主题
50	Central Warning System	中央警告系统
60	Central Display Systems	中央显示系统
32	**LANDING GEAR**	**起落架**
10	Main Gear and Doors	主起落架和舱门
20	Nose Gear & Doors, Tail Gear	前起落架和舱门，尾橇
30	Extension & Retraction	起落架的收、放
40	Wheel Assembly and Brakes	机轮组件和刹车
50	Steering	转弯
60	Position & Warning	位置及警告
70	Supplementary Gear	辅助起落架
80	Service Clearance	使用间隙
33	**LIGHTS**	**灯光**
10	Flight Compartment	驾驶舱
20	Passenger Compartment	客舱
30	Cargo & Service Compts.	货舱和勤务舱
40	Exterior	外部照明
50	Emergency Lighting	紧急照明
34	**NAVIGATION**	**导航**
10	Flight Environment Data	飞行环境数据
20	Attitude and Direction	姿态及指引
30	Landing & Taxi Aids	着陆和滑行设备
40	Independent Position Determining	自主位置判断
50	Dependent Position Determining	非自主位置判断
60	Flight Management Computing	飞行管理计算
35	**OXYGEN**	**氧气**
10	Crew	机组
20	Passenger	旅客
30	Portable	手提式
36	**PNEUMATIC**	**气源系统**
10	Distribution	分配
20	Indicating	指示
37	**VACUUM**	**真空系统**
10	Distribution	分配
20	Indicating	指示

CHAPTER 章	SUBJECT	主题
38	**WASTE/WATER**	废水 / 水
10	Potable	饮用水
20	Wash	冲洗
30	Waste Disposal	废水处理
40	Air Supply	气源
39	**ELECTRICAL/ELECTRONIC COMPONENTS AND MULTIFUNCTION UNITS**	电气 / 电子部件和多功能组件
49	**AIRBORNE AUXILIARY POWER**	机载辅助动力
10	Powerplant	动力装置
20	Engine	发动机
30	Engine Fuel and Control	发动机燃油和控制
40	Ignition/Starting	点火 / 起动
50	Air	空气
60	Engine Controls	发动机操纵
70	Indicating	指示
80	Exhaust	排气
90	Oil	滑油
51	**STRUCTURES/STANDARD PRACTICES**	结构 / 标准工艺
52	**DOORS**	门
10	Passenger/Crew	旅客 / 机组
20	Emergency Exit	紧急出口
30	Cargo/Baggage	货舱 / 行李舱
40	Service	服务设备
50	Fixed Interior	固定的内部装饰
60	Entrance Stairs	登机梯
70	Door Warning	门警告
80	Landing Gear	起落架
53	**FUSELAGE**	机身
10	Main Frames	主框架
20	Auxiliary Structures	辅助结构
30	Plate/Skin	板 / 蒙皮
40	Attach Fittings	接连接头
50	Aerodynamic Fairings	整流罩
60	Strakes	导流片

CHAPTER 章	SUBJECT	主题
54	**NACELLES/PYLONS**	**短舱 / 吊架**
10	Main Frame	主框架
20	Auxiliary Structures	辅助结构
30	Plates/Skin	板 / 蒙皮
40	Attach Fittings	连接接头
50	Fillets/Fairings	填角整流条 / 整流罩
55	**STABILIZERS**	**安定面**
10	Horizontal Stabilizers	水平安定面
20	Elevators	升降舵
30	Vertical Stabilizers	垂直安定面
40	Rudder	方向舵
50	Attach Fittings	连接接头
56	**WINDOWS**	**窗**
10	Flight Compartment	驾驶舱
20	Cabin	客舱
30	Door	门
40	Inspection & Observation	检查和观察
50	Flight Surfaces	飞行操纵面
60	**STANDARD PRACTICES-PROPELLER**	**标准施工—螺旋桨**
61	PROPELLERS	螺旋桨
70	**STANDARD PRACTICES-ENGINE**	**标准施工—发动机**
10	Cleaning	清洗
20	Inspection/Check	检验 / 检查
30	Correction/Rep.	校正 / 修理
40	Disassembly/Assembly	分解 / 装配
50	Materials/Hardware	材料 / 金属零件
60	Servicing/Preserving	勤务 / 油封
70	Testing	试验
80	Processes	工艺规程
90	Parts Manufacturing	零部件制造
71	**POWERPLANT GENERAL**	**动力装置概述**
10	Cowling	整流罩
20	Mounts	固定架
30	Fire seals	防火封严条

CHAPTER 章	SUBJECT	主题
40	Attach Fittings	连接接头
50	Electrical Harness	电气导线束
60	Air Intakes	进气
70	Engine Drains	发动机放沉淀
72	**ENGINE (TURBINE/TURBOPRO)**	**发动机（涡轮／涡桨）**
10	Reduction Gear and Shaft Section	减速齿轮及轴部段
20	Air Inlet Section	进气部段
30	Compressor Section	压气机部段
40	Combustion Section	燃烧部段
50	Turbine Section	涡轮部段
60	Accessory Drives	辅助传动
70	By-Pass Section	内外函部段
73	**ENGINE FUEL AND CONTROL**	**发动机燃油和控制**
10	Distribution	分配
20	Controlling	控制
30	Indicating	指示
74	**IGNITION**	**点火**
10	Elect. Power Supply	电源
20	Distribution	分配
30	Switching	转换
75	**AIR**	**空气**
10	Eng. Anti-Icing	发动机防冰
20	Accessory Cooling	附件冷却
30	Compressor Bleed Control	压气机引气控制
40	Indicating	指示
76	**ENGINE CONTROL**	**发动机控制**
10	Power Control	动力控制
20	Emergency Shutdown	紧急停车
77	**ENGINE INDICATING**	**发动机指示**
10	Power	动力
20	Temperature	温度
30	Analyzers	分析器
40	Integrated Engine Instrument Systems	综合发动机仪表系统
78	**EXHAUST**	**排气**

CHAPTER 章	SUBJECT	主题
10	Collector/Nozzle	集流环 / 喷口
20	Noise Suppressor	消音器
30	Thrust Reverser	反推
40	Supplementary Air	补充空气
79	**OIL**	**滑油**
10	Storage	仓库，贮存
20	Distribution	分配
30	Indication	指示
80	**STARTING**	**起动**
10	Cranking	手摇起动
81	**TURBINE**	**涡轮**
10	Power Recovery	功率恢复
20	Turbo-Supercharger	涡轮增压器
82	**WATER INJECTION**	**喷水**
83	**ACCESSORY GEAR BOXES**	**附件齿轮箱**
10	Drive Shaft Section	传动轴部段
20	Gear Box Section	齿轮箱部段
90	**CHARTS**	**图表**

Appendix II Glossary of Commonly-used Terms

English	Chinese
(A)Symmetrical	（不）对称
(In)Operative	（不）工作
(In)Sufficient	（不）充足
(Nm)Nautical mile	海里
A/D (Analog To Digital)	模数转换
A/THR	自动油门
Abnormal	不正常
Abrasion	磨损
Absence	没有，缺乏
AC (Alternating Current)	交流电
Accelerate	加速
Access	接近
Access Panel	接近盖板
Access Platform	工作梯
Accessory	附件
Accessory Gear Box	附件齿轮箱
Accommodate	调节，调和
Accomplish	完成
Accordance	符合
According to	根据
Accordingly	因此
Accumulator	蓄压器
Accuracy	精度
Achieve	完成，达到
Acid	酸
Acoustic	声音的
Acquire	获得

English	Chinese
Acquisition	获取
Activate	激活，接通
Active	主动地；活泼的
Active Mode	主动模式
Adapter	转接头
Additional Fuel Tank	附加油箱
ADF (Automatic Direction Finder)	自动定向机
Adhesive Film	胶膜
ADIRS (Air Data Intertial Reference System)	大气数据及惯导系统
ADIRU(Air Data Intertial Reference Unit)	大气数据惯导组件
Adjustable	可调的
Advance	发展，先进
Aerodynamic Contour	气动外形
Affect	影响
Affirmative	肯定的
After Cargo Heating	后货仓加热
Against	防止
Fire Agent	灭火器
AIDS (Aircraft Integrated Data System)	飞机综合数据系统
Aim	目的，瞄准
Air Blast	气流
Air Cycle Machine	空气循环机
Air extract	抽气
Air Stream	气流
Airborne	空中 / 机载的
Aircraft Electrical Control	飞机电传操纵
Aircraft Maint. Configuration	飞机维护构型
Airfield	机场
Airflow	气流
Airfoil	机翼
Airspeed (A/S)	空速
Airway	航路
AL Clad Sheet	包铝板
Alcohol	酒精

English	Chinese
Alert	警报
Alignment	定位，对准
Alphabetic	按字母排序的
Alternate	备用的；备用机场
Alternate Brake	备用刹车
Altimeter	高度表
Aluminium Alloy	铝合金
Amber	琥珀色
Ambient	环境
Ambient Light	环境灯光
Amount of	大量
Ampere	安培
Amplifier	放大器
Analog Signal	模拟信号
Analogue	模拟的
Angle Gearbox	伞齿轮箱
Angle Of Attack	迎角
Ann	安
Announcement	通知
Anode	阳极
Anti-Ice Valve	防冰活门
Antenna	天线
Anticipator	预选器
Anti-Fire Wall	防火墙
Anti-Icing	防冰
Anti-Skid	防滑
Antistatic	防静电的
AOA Probe	迎角探头
A/P(Autopilot)	自动驾驶
Appliance	器材
Applicable	适用的
Applicant	申请人
Approach	搓近
Approximate	大约

English	Chinese
Apron	裙板，围板
Aramidfiber	芳纶纤维
Arm	待命
Armature	电枢
Armed	待命的
Armrest	扶手
Arrow	箭头
Artificial Feel	人工感觉
Assist	帮助
Associated	关联的，相关的
Assume	假定
Asymmetrical	不对称的
ATC (Air Traffic Control)	空中交通管制
ATC Transponder	ATC 应答机
Atmosphere	环境，大气
Attach Fitting	接头
Attachment	接头
Attendant	乘务员
Attention	注意
Audio	音频
Aural	听觉的
ABS (Automatic Barking System)	自动刹车系统
Automatic Mode	自动模式
Automatic Test Procedure	自动测试程序
Available	可用的
Avionics	航空电子设备
Avionics Bay	电子舱
Avionics Compartment	电子舱
Avoid	避免
Axis	轴
Axial Flow Compressor	轴流式压气机
Back up	备用
Backrest	靠背
Baffle	隔板

English	Chinese
Baggage Bar	行李围栏
Bar	结构：棒料；电气：巴；机械：杆
Barometric Pressure	气压
Basic	基本
Beacon	信标
Beam	横梁
Bearing	电子：方位角；机械：轴承
Bellow	波纹管
Belly	机腹
Belly Fairing	机腹整流罩
Bend	弯曲
Bend Angle	弯曲角
Bend Radius	弯曲半径
Bezel	仪表前盖
Bird Impact (Strike)	鸟击
Blade	叶片
Blanking Plug	堵塞
Bleed Valve	发动机：放气活门；空调：引气活门
Blend Out	打磨去除
Blind Rivet	拉铆钉
Blockage	堵塞
Blocking Nut	防松螺帽
Blower	鼓风机
Board Level	电路板级
Bolt	螺栓
Bond	结构：粘接；电气：搭地
Bonding Strap	接地线，搭铁带
Boost Pump	增压泵
Bore	腔，孔
Borescope	孔探
Bowl	杯
Bracket	支架，接耳
Breather	通气装置
Bridge	电桥

English	Chinese
Brief	简洁的，短暂的
Brightness	亮度
Brush	刷子
Buffer	缓冲器
Buffet	餐具柜
Built-In Test	内置测试
Bulb	灯泡
Bullet	导向锥
Burst	爆裂
Bus	强电系统：汇流条；弱电系统：数据总线
Bushing	衬套
Butterfly-Type Valve	蝶形阀
Buzzer	蜂鸣器
By Means Of	通过……手段
Bypass Valve	旁通活门
Cabin Altitude	客舱高度
Cabin Attendant Seat	乘务员座椅
Cabin Depressurization	客舱失压
Cable	机械：钢索；电气：电缆
Calibrate	校准
Calibration	校验
Cancel	取消
Capacitor	电容
Captain	机长
Carbon	碳
Carbon Brake	碳片刹车
Carbonfiber	碳纤维
Card	板卡
Carpet	地毯
Carrier Injections	载波注入
Carry Out	实施，执行
Case	壳体，机匣
Cavity	洞，孔，腔
Ceiling	天花板

English	Chinese
Center	中心
Centering Spring	定中弹簧
Centrifugal Flow Compressor	离心式压气机
Certificate	证书，执照
Chamber	腔
Chamfer	倒角
Channel	信道
Charge	负责，充电（气，液）
Chassis	底架
Check Valve	单向活门
Chemical	化学的
Chemical Oxygen Generator	化学氧气发生器
Choke	扼流圈
Chronometer	计时器
Circuit	线路，电路
Circuit Breaker	电路断路器（跳开关）
Circumferential Joint	环向接缝
Circumstance	环境
Civil Transport	民用运输
Clad	金属包层
Clamp	夹子
Cleaning	清洗
Cleaning Agent	清洁剂
Clearance	间隙，许可
Clearview Window	活动窗，观察窗
Climb	爬升
Clip	夹子
Clipnut	夹板螺帽
Close	油路：关闭；电路：闭合，接通
Close-Up	结束工作
Clutter	杂乱
Coating	涂层、漆层
Coatroom	衣帽间
Coaxial Shaft	同心轴

English	Chinese
Code	编码
Coil	线圈
Cold Form	冷成形
Collar	螺帽
Combustible	易燃的
Combustion	燃烧
Command	命令
Common Nozzle	尾喷管
Comparator	比较器
Comparison	比较，对照
Compass	罗盘
Compensator	补偿器
Composite Material	复合材料
Compressed Air	增压空气
Compressor	压气机
Compressor Stall	压气机失速
Comprise	由……组成
Condenser	冷凝器
Conductor	导体
Cone	锥体
Configuration	构型
Confirm	证明，确认
Confirmed	确定的
Connector	机械：管接头；电气：电接头
Consider	考虑
Console	侧操纵台
Consumable	消耗性的
Consume	消耗
Contact	接触，接触器，触点
Container	集装箱
Contamination	污染
Continuously	持续的
Control Column	驾驶盘（杆）
Controller	控制器

English	Chinese
Conventional	传统的
Convert	转换
Copper	铜
Core	核心
Corrective Action	纠正措施
Corrosion	腐蚀
Corrosion Preventive Compound	防腐剂
Corrosion Resistant Steel	不锈钢
Cost	成本
Cotton Pin	开口销
Countersink Head	埋头
Cove Light Panel	拱形灯，光盖板
Cowl	整流罩
Crack	裂纹
Crack Stopper	止裂板
Crease	折痕
Crew	机组
Cross Beam	横梁
Cross Bleed	交叉引气
Cross Feed Valve	交输活门
Crossfeed	交输
Crossfeed System	交输系统
Cruise	巡航
Crystal	晶体
Current	电流
Cursor	游标
Curtain	隔帘
Curve	曲线
Customer Layout	客舱布局
Cutout	切口
Cutting Line	切割线
CVR (Cockpit Voice Recorder)	驾驶舱语音记录器
Cyan	宝石蓝
Cylinder	有气体的系统：气瓶；液压系统：作动筒

English	Chinese
Damage	损伤
Damping Mode	阻尼模式
Damping Orifice	阻尼孔
Dark Cockpit Philosophy	驾驶舱全黑逻辑
Data Base	数据库
DC (Direct Current)	直流电
Deactivate	解除
Debonding	脱胶
Debris	碎片，残骸
Deceleration	减速
Decimal	十进制的，小数点
Decode	译码
Decorative Panel	装饰板
Decrease	减少
Dedicated to	用于，专注于
De-Energize	断电
Defect	缺陷
Definition	定义
Deflate	放气
Defogging	除雾
Defuel	排油
De-Icing	除冰
Delaminate	分层
Delay	延时器，延误
Deliver	发送
Demand	要求
Demodulation	解调
Density	密度
Dent	凹坑
Departure	离开，离场
Depend on	取决于
Deployment	展开
Deposit	沉淀
Descend	下降

English	Chinese
Description	说明
Description And Operation	说明与操作
Design	设计
Designator	名称
Destination	目的地
Detail	细节
Detailed Visual Inspection	详细目视检查
Detect	探测
Determine	确定
Deviation	偏差
Device	装置
DFDR (Digital Flight Data Recorder)	数字式飞行数据记录器
Diagnosis	诊断
Diagram	图像
Difference	差值
Differential Linkage	差动连杆
Differential Pressure	压差
Differential Pressure Switch	压差电门
Diffuser	扩压器
Digital	数字的
Digital Data Bus	数字式数据总线
Diode	二极管
Dipped in	浸染
Direct Current	直流
Directly	直接的
Discard	报废
Discharge	排放，释放
Discharger Brush	放电刷
Discolor	变色
Disconnect	脱开
Discrepancy	不符合，差误
Discrete	离散的
Discriminator	鉴频器
Disk	盘

English	Chinese
Disregard	忽略，忽视
Distortion	变形
Distribution	分配
Ditching	水上迫降
DME (Distance Measuring Equipment)	测距机
Door Frame	门框
Dorsal Fin	背鳍
Dot	点
Doubler	加强板
Downstream	下游
Drag Strut	承拉支柱
Drain	余油，排水
Drawing	图纸
Dress Out	（凹坑）整形
Drift	漂移
Dry Cranking	干转
Dual	二重的
Duct	函道
Duct Velocity	管道流速
Due to	由于
Duplex	双路
Dust-Free	无尘
Dynamic	动态的，有活力的
Eddy Current	涡流
Edge Distance	（紧固件）边距
Effectivity	有效性
EFIS (Electronic Flight Instrument System)	电子飞行仪表系统
Ejector	引射口
Electric	电气的
Electric Motor	电动马达
Electric Pump	电动泵
Electric Wiper	电动雨刷
Electrical Bonding	电搭接
Electrical Heating	电加温

English	Chinese
Electrical Network	电网
Electrical Wire	电导线
Electrically Driven Valves	电动活门
Electrically Operated	电控
Electro-Hydraulic Valve	电控液动活门
Electromotor	电动机
Electronic	电子的
Electro-Pneumatic Valves	电控气动活门
Electrostatic	静电
Element	成分，要素，元件
EMER GEN	液压驱动应急发电机
Emergency	紧急情况
Emergency Descent Device	应急下降装置
Emergency Equipment	应急设备
ELT (Emergency Locator Transmitter)	应急定位信标发射机
Encode	编码
Energize	接通，通电
Engine Driven Generator	发动机驱动发电机
Engine Driven Pump	发动机驱动泵
Engine Fan	发动机风扇
Enhance	提高，加强
Entire	全部的
Entrance=Entry=Intake	入口
Envelop	飞行包线
Environment	环境
ECS (Environmental Control System)	环境控制系统
Epoxy Adhesive	环氧树脂
Equivalent	等效
Equivalent Resistor	等效电阻
Erosion	侵蚀，风蚀
Error	错误
Escape	流出，逃跑
Escape Rope	撤离绳
Escape Slide	撤离滑梯

English	Chinese
Essential Bus	主汇流条
Establish	建立
Estimate	评估，估计
Etch	侵蚀，蚀刻
Evacuate	疏散
Evacuation	应急撤离
Even	即使；偶数
Examine	观察
Exchange	交换
Exciter	励磁，励磁机
Exhaust Cone	排气锥
Exhaust Gas Temperature	排气温度
Extend /Extension	伸出，放下
Extension Board	扩展电路板
Extinguish	熄灭
Extinguishing Agent	灭火剂
Extracted	抽出
Extraction Duct	抽气管道
Extraction Fan	抽气风扇
Extrusion	挤压型材
FAC (Flight Augmentation Computer)	飞行增稳计算机
Fail	失效
Fairing	整流罩
Familiar	熟悉的
Familiarize	使熟悉
Fan	风扇
Fan case	风扇机匣
Fan Cowl	风扇罩
Fastener	紧固件
Fastener Spacing/Pitch	紧固件间距
Fatigue	疲劳
FD (Flight Director)	飞行指引针
Feedback	反馈
Feeder	馈线

English	Chinese
Fence	翼刀，栅栏
Fiberglass	玻璃纤维
Field	磁场
Field Current	励磁电流
Field Elevation	机场标高
Figure	示图
Filler	填片
Fillet Seal	填角密封
Filter	过滤器
Filter Cartridge	滤筒
Filter Element	滤芯
Finger Doubler	指形板
Fire Wall	防火墙
First Officer	副驾驶
Fits And Clearances	配合与间隙
Fixed Window	固定窗
Fixture	夹具
Flag	故障旗
Flammable	可燃的
Flange	凸缘，法兰盘，安装边
Flap	飞行操纵：襟翼；其他：档板
Flashlight	手电筒
Flexible Bellow	柔性波纹管
Flexible Hose	软管
Flight Control Surface	飞行操纵面
Flight Path Angle	飞行航迹角
Flight Phase	飞行阶段
Flight profile	飞行剖面
Floor Covering	地板覆层
Floor Structure	地板结构
Flow Rate	气流流量
Flow Restrictor	限流器
Flow Transmitter	流量传感器
Flowchart	流程图

English	Chinese
Flowmeter	流量计
Fluorescent Lamp	荧光灯
Flush	冲洗
Fly-By-Wire	线控 / 电传操纵飞行
FOB (Fuel On Board)	机载燃油重量
Folding Seat	可折叠座椅
Following	以下的
Food Tray	餐桌板
Forging	锻件
Formed Section	板弯型材
Frame	框架
Framework	框架
Free Fall Extension	重力放轮
Freon	氟利昂
Frequency	频率
Fresh Air	新鲜空气
Fresh Water	淡水
Friction	摩擦
Fuel Flow	燃油流量
Fuel Pump	燃油泵
Fuel Quantity	燃油量
Fume	烟雾
Furnishing	客舱设备
Fuse	熔丝
Fusible Plug	易熔塞
Gage	表，量块
Gage Pressure	表压
Galley	厨房
Gas Turbine Engine	燃气涡轮发动机
Gasket	衬垫，密封垫
Gauge	量表
Gear	A 起落架；B 齿轮；C 工具
General Visual Inspection	一般目视检查
Generate	产生

English	Chinese
Generator	发电机
Gimbal	平衡框，万向节
Glareshield	遮光板
Glove	手套
Goggle	护目镜
Gouge	凿痕
Governor	调速器
Gravity	重力
Grip Length	夹紧长度
Groove	槽
Ground Crew	地面人员
Guarded	带保护盖的
Guide	导向块
Gust Lock	防风锁
Gyro	陀螺
Halogen	卤素
Hand Pump	手摇泵
Handset	手持话筒
hangar	机库
Hardness	硬度
Harmonic	谐波
Harness	导线束
Hazardous	有害的
Heading	航向
Headset	耳机
Heat Exchanger	热交换器
Heat Sink	散热器
Heat Treatment	热处理
Hence	从此时起，因此
Hexagonal Wrench	六角扳手
High Bypass Rate	高涵道比
Hi-Lok	高锁钉，海螺钉
Hinge	铰铸
Hoist	吊车

English	Chinese
Honeycomb Core	蜂窝夹芯
Housing	壳体
Hydraulic Circuit	液压回路
Hydraulic Reservoir	液压油箱
Hydrogen-Charged	充氢的
Hydro-Mechanical	液压机械式
Icing Condition	结冰条件
Identical	相同的
Identification Label	识别标签
Identify	鉴别
Illuminate	照明，照亮
Illustration	图示
Impact	碰撞，坠落
Impedance	阻抗
Impeller	叶轮
Imperative	强制性的
In Accordance With	根据
In Parallel	并联
In Series	串联
In Turn	依次
Inadvertent	不注意的
Incandescent Light	白炽灯
Inclinometer	倾度计
Indentation	缺口
Independent	独立的
Index	索引
Inductor	电感
Inflation	充气
Ingestion	摄取
Inhale	吸入
Inhibited	抑制的
Initiate	开始，启动
Injury	损伤
Inlet	入口

English	Chinese
Inlet Cowl	进气道前环
Inlet Guide Vane	进口导向叶片
Inop=Fails	失效
Inoperative	不工作
Input	输入
Insert	v. 插入；n. 钢丝螺套
Inspection	检查，检验
Instruction	说明，说明书
Instrument	仪表
Insufficient	不足的
Insulation	隔离层
Insulation Blanket	隔离毡
Intake	进气
Integral Tank	整体油箱
Intensity	亮度
Intercell	内组油箱
Intercom	内话
Interconnect	互联
Interface	机械：接触面；电子：接口
Interface Unit	接口组件
Interfere	干扰
Interlock	互锁
Interphone	内话
Inter-Relationship	内部联系
Interrogate	询问
Interrupt	中断
Interval	间隔
Intervention	介入，干涉
Invalid	无效的
Investigate	调查，研究
Ionize	电离
Isolation	绝缘
Jack	千斤顶
Jet Pump	引射泵

English	Chinese
Job Set-Up	工作准备
Joint	连接，接头
Keel Beam	龙骨梁
Key	键
Kit	工具包，组件
Knob	旋钮
Knot	节（海里 / 小时）
Label	标签
Labyrinth Seal	篦齿型（迷宫型）密封
Laminated Structure	分层结构
Landing Elevation	着陆机场高度
Landing Elevation Selector	着陆标高选择器
Landing Gear Bay	起落架舱
Landing Gear Gravity Extension	重力放轮
Lane	通道
Lapping	研磨
Lateral	横向的
Latitude	纬度
Lavatory	厕所
Layout	布局
Leading Edge	前缘
Leakage	渗漏
Legend	字符灯
Life Vest	救生衣
Light Alloy	轻质合金
Line Key	行选键
Liner	衬垫
Lining	衬层，装饰板
Liquid Crystal Display (LCD)	液晶显示器
Load	载荷
Load Board	负载电路板
Load Compressor	负载压气机
Localizer	定位信标，航道
Location	位置

English	Chinese
Locator	定位器
Lock Wire	保险丝
Lockbolt	自锁螺栓
Log Book	履历本
Logic	逻辑
Longeron	纵梁
Longitude	经度
Longitudinal Joint	纵向接缝
Loop	环路
Lug	接耳
Mach Number	马赫数
Machanism	机械装置
Magnet Wheel	磁轮
Magnet	磁性
Magnetic Chip Detector	磁性碎屑探测器（磁堵）
Main Gear Bay	主轮舱
Maintain	维持
Major Alternation	重要改装
Major Repair	重要修理
Majority	大部分，多数
Male Connectors	公插头
Malfunction	故障
Mandatory	强制的
Maneuver Speed	机动速度
Manifold	总管
Manual	人工的，手册
Manual Mode	人工模式
Manufacturer	制造商
Margin	（紧固件）边距
Mass Flow	质量流量
Measure	测量
Melt	融化
Memo	信息，提示
Mend	修补

English	Chinese
Metering Valve	计量活门
Metric Unit	米（公）制单位
Micro-Ammeter	毫安表
Micro-Switch	微动电门
Millibar	毫巴
Minor Alternation	次要改装
Minor Repair	次要修理
Miscellaneous	混杂的，杂项
Missed Approach	复飞
Mixing Unit	混合器
Mode Selector-Valve	模式选择活门
Modify	改装
Modulation	调制
Module	模块，组件
Moisture	水汽
Moisture-Free	无水汽
Monitor	监控，监控器
Motor	马达
Mount	安装，固定
Muffler	消音器
Multichannel	多道的
Multi-Meter	万用表
Multipurpose	多用途
MUX (Multiplexer)	混频器
Nacelle	短舱
Nautical Mile (NM)	海里
Navaid	助航系统，导航台
Negative	负，否定的
Network	网络，电路，电网
Nick	缝隙
Nitrogen	氮气
NO GO	不合格
Non Return Valve	单向活门
Nose	机头

English	Chinese
Nose Cowl	进气道前环
Nosewheel Steering System	前轮转弯系统
Noxious	有害的
Nozzle	喷嘴
Nut	螺帽
Nutplate	托板螺帽
Nylon	尼龙
Observe	遵照
Observer	观察员
Obstacle	障碍
Obstruction	堵塞
Obtain	得到，获得
Occupant	乘员
Occur	发生，举行
Ohm	欧姆
Oil Separator	滑油分离器
Oil Strainer	滑油筛
Oil-Pneumatic	油—气混合式
On Board	飞机上
Open	油路：打开；电路：断开
Open Flame	明火
Operational Check	操作测试
Operative	工作
Optical	光学的
Optimize	优化
Optional	选装件；选装的
Orient	定向
Orifice	量孔，限流孔
Oscillator	振荡器
Oscilloscope	示波器
Out Board	外侧
Out Of Limit	超限
Outer Cell	外组油箱
Outflow Valve	排气活门，溢流活门

English	Chinese
Outlet	出口
Output	输出
Outside Diameter	外径
Oven	烘箱
Over Wing Emergency Exits	翼上应急出口
Overall	全面
Overboard	机外
Overcoat	保护层
Overflow Valve	有液体的系统：溢流活门；有气体的系统：放气活门
Overhead Panel	头顶面板
Overhead Stowage Compartment	行李架
Overheat	过热
Override	超控
Overwing	翼上的
OVRD (Override)	超控
PA (Passenger Address)	旅客广播
Packaging	包装
Packing	衬垫，包装
Pad	填塞
Parallel	平行的
Parameter	参数
Parking Brake	停留刹车
Particular	特有的
Partition Wall	隔板
Parts & Material	零件和航材
Passenger Service Unit	旅客服务组件
Patch	补片
Pattern	图形
Pay Load	有效负载
PB (Push Button)	按钮
Pedal	踏板
Pedestal	操纵台
Pendulum	摆
Perforate	打孔

English	Chinese
Performance	性能
Permanent	永久的
Permit	允许
Personnel	人员
Phase	阶段，相，相位；时期
Philosophy	哲学，理论
Photodiode	光敏二极管
Photoelectrical	光电的
Photo-Luminescent Strip	发光带，荧光带
Piezo-Resistive	压阻的
Pin	机械：销钉；电气：插针
Pink	粉红色
Pitch	俯仰
Pitch Trim Wheel	俯仰配平手轮
Pitot Tube	皮托管
Placard	标牌
Plastic Disposal Bag	一次性塑料袋
Plate	板材（≥ 0.25 英寸）
Plenum	整流腔
Plier	钳子
Plug	堵头
Pneumatic	气源，气动的
Pointer	指针
Polarize	极化
Pop Out Indicator	弹出式指示器
Port	口
Portable	便携的
Portable Fire Extinguisher	手提式灭火瓶
Position	位置
Positive	正
Postire	正的
Potable Water	饮用水
Potential	潜在的
Potentiometer	电位计

English	Chinese
Pouch	储存袋
Power Lever Angle	油门杆角度
Power-Up Test	上电测试
Pre	在……之前
Precaution	预防措施
Pre-Cooler	预冷器
Prediction	预测
Preparation	准备，预备
Preselect	预选
Presence	在位
Presentation	介绍，陈述
Pressure Bulkhead	承压隔框
Pressurization	增压
Prevent	防止
Primary	首要的，主要的
Primary Structure	主要结构
Primer	底漆
Prior	优先的
Priority	优先权
Priority Over	优先于
Probe	探头
Procedure	程序
Process The Data	处理数据
Processing	处理
Protruding Head	凸头
Proximity	接近
Proximity Sensor	接近传感器
PTU (Power Transfer Unit)	动力转换组件
Puller	拔具
Pulley	滑轮
Pulse	脉冲
Pure Oxygen	纯氧
Purpose	用途，目的
Purser	乘务长

English	Chinese
Pushbutton	按钮
Pushbutton Switch	按钮电门
Pylon	发动机吊架
Quality	质量
Quantity	数量
Quick Release Pin	快卸钉
Radar	雷达
Radio Altimeter	无线电高度表
Radius	半径
Radome	雷达罩
Raft	救生船
Rain-Repellent	排雨刷
Ram Air	冲压空气
RAT(Ram Air Turbine)	冲压空气涡轮
Ratio	比率
Readout	读数
Rear	后部
Receptacle	插座
Recharge	充（气，电）
Recirculated Air	再循环空气
Recirculation	再循环
Recline	后靠
Rectification	校正
Rectifier	整流器
Reduced Thrust	减推力
Redundancy	冗余
Reference	参照
Refrigeration	制冷
Refrigerator	冰箱
Refuel	加油
Regulate	调节
Regulate Valve	调节活门
Regulator	调节器
Reject	报废

English	Chinese
Relate	联系
Relay	继电器
Release	释放，松开
Relief Valve	释压活门
Remain	保持
Remark	备注
Repair Angle	修理角材
Representative	代表
Reserved Thrust	储备推力
Residual	多余的，残留的
Resin	树脂
Resin And Glass Fiber Laminate	树脂和玻璃纤维层
Resistance	电阻
Resistor	电阻
Resolution	分辨率
Resolver	解算器
Resonance	谐振
Respective	各自的
Restore	再现，复原，存储
Restrictor	限流器
Restrictor Check Valve	限流单向活门
Retainer	保持块
Retract Forward	向前收起
Retract Inboard	向内收起
Reverse Flow	回流，反流
Reverser	反向器
Revert	恢复
Revert to	转到
Review	回顾，检查
Rib	肋板
Rig Pin	校装销
Rinse	冲洗
Rivet	铆钉
Rod	杆

English	Chinese
Roll Out	滑跑
Roller	滚棒
Root	根（底）部
Rotable Part	周转件
Rotation	抬前轮；旋转
Rotor	转子
Routed	已选择路径
Runway	跑道
Safety Collar	保护环
Safety Goggles	护目镜
Sander	打磨器
Sandwich Panel	夹芯板
Satcom System	卫星通信系统
Scale	刻度，刻度表，秤
Scan	扫描，浏览，细看
Scarf Sanding	斜面打磨
Scavenge	回油
Schematic	图解的
Scratch	划痕
Screen	显示屏，滤网，格栅
Screw	螺钉
Seal	密封
Sealant	密封剂
Sealed Vapor Lamp	密封蒸汽灯
Seat Belt	座椅安全带
Seat Cushion	坐垫
Seat Rail/Track	座椅导轨
Secondary Structure	次要结构
Section	部分，章节，截面
Segment	部分，段
Selector	选择器
Selector Switch	选择电门
Self Monitoring Function	自监控功能
Self-locking	自锁

English	Chinese
Self-Test	自测
Semiconductor	半导体
Sensing Element	敏感元件
Sensitive	敏感
Sensitivity	灵敏度
Sensor	传感器
Sensor Board	传感器电路板
Separate	单独的；隔离
Sequence	顺序
Serviceability	可维护性，有用性
Servo-Control	伺服控制系统
Servo-Valve	伺服活门
Setscrew	机械：定位螺钉；仪表：调节螺钉
Shear Head	剪切头
Sheet	薄板（<0.25 英寸）
Sheet Thickness	板厚
Shield	屏蔽
Shim	填隙片
Shock Absorber	减震器
Shock Proof	防震
Shock Strut	减震支柱
Short	短路
Shroud	防护罩
Shut Down	关闭
Shut Off	关断
Shutoff Valve	关断活门
Side Cowl	侧整流罩
Side Stick	侧操纵杆
Sidewall Panel	侧壁板
Silicone	硅树脂
Simplify	简化
Simulate	模拟
Simulator	模拟器
Simultaneously	同时的

English	Chinese
Single Side Band	单边带
Single/Dual Lane Slide	单 / 双通道滑梯
Single-Aisle	单通道
Situation	情况
Skin	蒙皮
Slave	从动器 / 伺服器
Sleeve	套管，衬套
Slide Raft	滑梯式救生船
Sliding Window	活动窗
Smoke Detector	烟雾探测器
Smoke Hood	防烟面罩
Snap	卡簧
Snapshot	快照，短暂
Soft Brush	软毛刷
Software	软件
Solder	焊料，焊接
Solenoid	线圈，电磁阀
Solid Rivet	实芯铆钉
Solvent	溶剂
Sound Suppression Liner	消音衬
Source	源头
Spacer	间隔片，垫片
Spanner	扳手
Spar	翼梁
Spar Box	梁盒
Spark Proof Lamp	防火花照明灯
Special	特殊
Special Tools, Fixtures & Equipment	特殊工具，工装和设备
Specification	规范
Speed Brakes	减速板
Spin	旋转
Splice Strap	搭接带
Spline	花键
Split	分开

English	Chinese
Spoiler	扰流板
Spray	喷射
Spring	弹簧
Squib	爆炸帽
Stability	稳定性
Stabilize	稳定
Stable	稳定的
Stage	级
Stainless Steel	不锈钢
Stair Bay	客梯舱
Stairway	客梯
Stanchion	支柱
Standby	备用
Start Lever	起动手柄
Starter	起动器
State	情形，状态
Static Discharge	静电放电
Static Inverter	静变流机
Static Port	静压口
Static-Dissipative	静电耗散
Station	站位
Stator	定子
Steady	稳定的
Stem	杆
Step Sanding	台阶打磨
Stop Hole	止裂孔
Stopwatch	秒表
Stowage	储藏间
Strictly	严格地
Stringer	长桁
Stripped Areas	褪漆区域
Substitute	替代
Subsystem	子系统
Suction	吸入

English	Chinese
Sufficient	足够的
Summarize	总结
Sump	油槽
Supervisor	监督人，主管
Supply Board	电源板
Support	支撑
Suppressor	抑制器
Surge	震动，喘振
Surroundings	环境
Swap	交换
Swapped Part	串件，对调件
Switch	名词：开关；动词：转换
Swivel	旋转
Symbol	标志，符号
Symmetrical	对称
Synchro Transmitter	同步发射器
Synchronizer	同步器
Tachometer Generator	测速发电机
Tag	挂牌
Tail Cone	尾锥
Take Off	起飞
Take Over	接管
Tape	磁带，胶带
Target	目标
Task	任务
Taxi	滑行
Tee	T字接头
Teflon	特氟隆
Telescopic Duct	伸缩管道
Temper	回火
Temperature Indication Strip	温度指示带
Temporary Revision	临时修改版次
Temporay	暂时的
Tension Head	拉伸头

English	Chinese
Terminal	接线柱，终端
Test Bench	测试台
Test Set	测试装置
Testing And Fault Isolation	测试与故障隔离
Thermal	热量的
Thermal Stripper	热脱焊机
Thermal SW	热敏电门
Thermocouple	热电偶
Thermostat	恒温器
Thickness	厚度
Thinner	稀释剂
Thread	螺纹
Threshold	门限，首检，开端，极限
Thrust	推力
Thrust Lever	推力杆
Thrust Reverser T/R	反推
Tie Down Point	系留点
Tire	轮胎
Titanium Alloy	钛合金
Titanium Plate	钛板
To Be Defined	待定
Toggle	轴节
Tolerance	公差
Topcoat	面漆
Torque	拧紧
Torque Motor	扭力马达
Torque Tube	扭力管
Torque Wrench	力矩扳手
Touch Up	补漆
Toxic	有毒的
Trace	航迹
Trailing Edge	后缘
Transceiver	收发机
Transducer	传感器

English	Chinese
Transfer Valve	传输活门
Transformer	变压器
Transformer Rectifier	变压整流器
Transistor	晶体管，三极管
Transit	转换
Translating Cowl	移动整流罩
Transmission	传导
Transmitter	传感器，发射机
Transponder	应答机
Travel Limitation Unit	行程限制组件
Travel Range	行程范围
Treatment	处理
Trigger	触发
Trim	配平
Trim Control	配平控制
Trip	跳开
Tube	管料
Tubeless Tyre	无内胎轮胎
Tubular	管状的
Tune	调谐
Turbine	涡轮
Turbine-Air Motor	空气涡流发动机
Turbofan	涡轮风扇
Turn	转弯
Turn Coordination	协调转弯
Twin Engine	双发的
Uf	欠频
Unavailable	不可用的
Unclad/Bare	非金属包层
Unpressurized Area	非增压
Upstream	上游
US Unit	英制单位
Uv	欠压
Vacuum	真空

English	Chinese
Vacuum Cleaner	吸尘器
Validity	有效性，合法性
Variation	变量
Varied	多样的
Various	多方面
Varnish Stripping	褪漆
Vendor	供应商
Vent	通风
Ventilation	通气
Venturi	文氏管
Verify	审核，确认
Version	版本
Vertical Stabilizer	垂直安定面
Vertical Strut	立柱
Via	通过
Vibration	振动
Vicinity	周围，附近
Viscosity	粘性
Visibility	视野
Visual	目视
Volt	伏特
Voltage	电压
Volume	音量，体积，卷
Walk Round	绕机
Warm Up	暖机
Warning Notice	警告牌
Washer	垫片
Waste Bin	废纸箱
Water Separator	水分离器
Waveform	波形
Way Point	航路点
Weather Radar	气象雷达
Web	腹板
Weld	焊接

English	Chinese
Wind Shear	风切变
Winding	绕组
Windshield	风挡
Wing Center Box	中央翼盒
Wing Root	翼根
Wingtip	翼尖
Wiper	风挡雨刷
Wiring	线路
Workbench	工作台
Workload	工作负荷
Wrap	包，捆
Wrench	扳手
Yoke	轭架
Zone	区域

References

1. 白杰，张帆．民航机务英语教程 [M]．北京：中国民航出版社，1997．

2. 李永平，魏鹏程．民航机务专业英语 [M]．北京：国防工业出版社，2014．

3. 刘建英．电子技术基础（ME）[M]．2 版．北京：清华大学出版社，2016．

4. 蒋陵平．燃气涡轮发动机（ME-TA，TH）[M]．2 版．北京：清华大学出版社，2016．

5. 屈静，于燕红，张建荣．民航机务职业英语口语 [M]．北京：机械工业出版社，2010．

6. 张铁纯．涡轮发动机飞机结构与系统（ME-TA）（上）[M]．2 版．北京：清华大学出版社，2017．

7. 任仁良．涡轮发动机飞机结构与系统（ME-TA）（下）[M]．2 版．北京：清华大学出版社，2017．

8. 任仁良．涡轮发动机飞机结构与系统（AV）（下）[M]．2 版．北京：清华大学出版社，2017．

9. 张鹏．涡轮发动机飞机结构与系统（AV）（上）[M]．2 版．北京：清华大学出版社，2017．

10. Boeing AMM[Z]．Boeing Company, 2001．

11. https://wenku.baidu.com/view/afa3e9c1aa00b52acfc7ca76.html

12. http://www.aero.cn/2011/0329/22530.html

13. http://www.b737.org.uk/powerplant.htm

14. http://www.360doc.com/content/15/1208/12/29514199_518725388.shtml

15. http://www.b737.org.uk/apu.htm